A New Understanding of Mental Disorders

A New Understanding of Mental Disorders

Computational Models for Dimensional Psychiatry

Andreas Heinz

The MIT Press
Cambridge, Massachusetts
London, England

This book was set in Stone Serif by Westchester Book Composition. Printed and bound in the United States of America.

Library of Congress Cataloging-in-Publication Data

Names: Heinz, Andreas, 1960– author.
Title: A new understanding of mental disorders : computational models for
    dimensional psychiatry / Andreas Heinz.
Description: Cambridge, MA : The MIT Press, [2017] | Includes bibliographical
    references and index.
Identifiers: LCCN 2017008084 | ISBN 9780262036894 (hardcover : alk. paper)
Subjects: | MESH: Mental Disorders—psychology | Mental
    Disorders—physiopathology | Models, Psychological | Nervous System
    Physiological Phenomena
Classification: LCC RC467 | NLM WM 140 | DDC 616.89—dc23
    LC record available at https://lccn.loc.gov/2017008084

10 9 8 7 6 5 4 3 2 1

Contents

Preface

The times they are a-changin'
—Bob Dylan

Research on the neurobiological correlates of mental disorders has seen a dramatic increase in both the scope and depth of investigation. Nevertheless, to date, many mental disorders are still classified according to criteria that solely rely on the manifestation of clinical symptoms and their change over time. To some degree, this situation is comparable to the state of knowledge with respect to "stroke" before cranial computed tomography was introduced into routine clinical diagnosis more than 40 years ago. With the help of such imaging methods, ischemic versus hemorrhagic strokes could easily be distinguished, and treatment was adjusted accordingly. Similar hopes were raised when functional magnetic resonance imaging (fMRI) started to be widely used in neurobiological research on mental disorders. Today, there is an abundance of findings relating certain cognitive experiences to their respective activation patterns. However, the impact of such imaging technics on clinical diagnosis and treatment with respect to most mental disorders was rather limited. Accordingly, two conclusions can and have been drawn.

The first one questions the validity of current diagnostic categories and suggests that neurobiological findings do not help to guide treatment decisions because these diagnostic categories rely on inadequate classifications of mental disorders. Accordingly, this dimensional approach suggests not to start with traditional diagnostic classifications (e.g., the distinction between schizophrenia and bipolar disorder) and then look for neurobiological correlates that clearly distinguish these two disease categories from each other. Instead, research should focus on the neurobiological correlates of key cognitive mechanisms (including working memory performance) and assess their various degrees of alteration across established nosological boundaries.

A second, not mutually exclusive approach suggests that our current and most widely used imaging techniques are not sophisticated enough to capture the relevant neurobiological alterations in mental disorders. This computational approach suggests that beyond identifying activation patterns that are associated with certain mental operations, mental operations themselves should be analyzed more diligently. Specifically, computational models should be generated that reflect decision-making processes and reveal key computational steps guiding behavior, which can then be used to search for biological correlates of these computational operations. A famous case in point is computation of an error in reward prediction, which drives reinforcement-based learning and has been associated with the amount of phasic dopamine release in the ventral striatum and related brain areas.

This book focuses on both approaches and tries to link them with a new look at the classification of mental disorders. Our approach supports the idea that key mechanisms of learning and decision making including Pavlovian conditioning as well as model-based and model-free instrumental behavior should be computationally modeled and assessed in different mental disorders, thus following a dimensional approach instead of limiting research to one traditionally defined disease category at a time. Furthermore, we suggest that current clinical classifications have become too complex and tend to label common states of human suffering as disorders, thus failing to focus on severe diseases including dementia, addiction, as well as major affective and psychotic disorders. We suggest that traditional disease categories have their clinical value, yet nevertheless are in need of critical reflection including a foundation in a philosophical anthropology, which reflects and respects the diversity of human experiences and limits itself to describing key mechanisms required for individual survival and living in a shared world with other human beings. It is to patients who experience severe mental disorders and display impressive creativity when trying to cope with their challenges and their therapists that this book dedicated.

Acknowledgments

This work would not have been possible without my academic teachers, Prof. Dr. Horst Przuntek, who introduced me to neurology and accompanied my professional development for more than a decade; Prof. Dr. Hanfried Helmchen, whose focus on ethical issues in psychiatry is inspiring psychiatrists to this day; Dr. Daniel R. Weinberger and Dr. Markku Linnoila, who shared their profound experiences with me during a stay at the National Institutes of Health; Prof. Dr. Fritz Henn and Prof. Dr. Karl Mann, who accompanied my professional development during work at the University of Heidelberg; and Prof. Dr. Axel Honneth and Prof. Dr. Hans-Peter Krüger, my philosophical teachers. This work would not have been possible without grant support provided by the European Union, the Deutsche Forschungsgemeinschaft (DFG), and the German Ministry for Research (BMBF). I am deeply indebted to a large number of friends and coworkers whose main publications are quoted within the following pages. I would also like to thank my family for their continual support and last but not least give my special thanks to all the patients and probands who participated in our studies. Experiencing mental disorders can cause tremendous suffering but also provide an unusual and therefore very important perspective on human life. Patients with depression are known to be more realistic and rational than nondepressed subjects (who generally tend to overestimate their individual importance), subjects with psychosis experience the fragility of our common understanding of human interactions, and patients with addictions encounter human desire in its deepest degree. All these experiences are not only deeply human but also provide alternative and often very creative views on our lifeworld, and while this book focuses on mechanisms trying to explain mental disorders, the contribution of such individual experiences to a deeper understanding of human life cannot be underestimated.

1 Introduction

Psychiatric research is impressively successful: neurocircuits have been identified that are activated when humans are confronted with affectively positive or negative stimuli; it has been elucidated how these neurocircuits are modulated by neurotransmitters such as dopamine and serotonin; genetic and environmental effects have been described that modulate such neurotransmitter systems; and these insights have helped us to better understand the effects of medication, as well as some psychotherapeutic interventions (Meyer-Lindenberg, 2010; Heinz et al., 2011; Wang and Krystal, 2014). Moreover, basic research has helped to identify the exact computational roles that certain neurotransmitters such as dopamine play in modifying behavior, and the same mathematical tools can be used to formalize empirical accounts of individual variability, providing us for the first time in the history of psychiatric research with the possibility to use the same computational approaches to describe behavior and its underlying neurobiology (Corlett and Fletcher, 2014; Friston et al., 2014). Instead of trying to correlate mood states as reported by individual subjects, such computational approaches can thus directly associate mathematical models of individual behavior with individual brain states.

Psychiatric research is in crisis. While current classifications of mental disorders have multiplied and sparked a highly controversial debate on whether all such patterns of behavior actually constitute mental disorders, let alone diseases (Frances and Raven, 2013), major research agencies such as the National Institutes of Health have stopped supporting research that is oriented at such traditional disease classifications and instead suggest to focus on basic dimensions of mental disorders such as reward learning or working memory. Such basic dimensions of mental disorders are supposedly not confined to specific traditional disorders. Instead, they should constitute new research domains that play a role in the development of a multitude of illness expressions and stages. Proponents of this approach claim that future research along these lines will require disease concepts and nosological classifications to be fundamentally reconceptualized (Insel et al., 2010).

There appears to be hardly any common ground between proponents of traditional disease classification systems on the one hand and researchers focusing on basic neurobiological mechanisms on the other: the first ones criticize that besides disorders such as dementia, neurobiological research has largely failed to help classify mental disorders in clinical situations; the latter claim that this failure is exactly due to an outdated way of looking at mental disorders. They suggest ceasing to classify mental disorders by key symptoms that have been handed down by generations of clinicians and instead suggest focusing on quantifiable models of human behavior and its underlying basic computational mechanisms.

Situations in which traditional certainties are questioned by modern research and in which the ensuing controversies appear to be fundamentally unsolvable have repeatedly been described in theoretical accounts of scientific development. For example, Kuhn (1962) suggested that "scientific revolutions" take place once established theories and the key paradigms on which they are built appear to be outdated: this is often the case when established paradigms have served their purpose; have helped to explain phenomena as best as they can; and scientists are becoming increasingly dissatisfied with the limits imposed by the established approaches to gaining further insights. Hence, new theories, which in many ways may be less elaborated than the traditional ones, will be tested, and scientists will start to use them exactly because they offer new insights, even if—at the present moment—they are overall less elaborate and useful than the traditional ones. In this respect, Fleck (1979) suggested that a key process in science is indeed learning by doing: by applying concepts and testing their usefulness, we learn to see the world through the eyes of such models; we understand where certain paradigmatic approaches and explanations can help, where they have to be refined, and where they are actually useless.

Thus, psychiatric research could be just at the point of such a scientific revolution—traditional concepts (although, as we will discuss, clinically useful and to date irreplaceable—will in the short or long run be replaced by new ways of looking at mental disorders. One of these ways is described in this book: a focus on basic learning mechanisms, which can be operationalized in computational models in order to better explain the development and maintenance of mental disorders. Basic learning mechanisms and the role they play in human behavior include Pavlovian conditioning; that is, the effects of Pavlovian conditioned cues on behavior (e.g., the effects of a bell, which always rings when food is arriving, on saliva production in Pavlov's dog). They also include operant learning from feedback such as reward and punishment and the effects of Pavlovian conditioned stimuli on such instrumental, goal-directed behavior. Such basic learning mechanisms can help to describe how goal-directed behavior can be transformed with increasing practice into habits that are hard to modify by new rewards or conscious decisions (imagine how hard it can be to change a well-established but hurtful move in sports or dancing).

Focusing on such learning mechanisms has two advantages: first, it directly refers to key traditions in animal research and behavior therapy; namely, the study of behavioral effects of reward and punishment as well as the association between cues and reactions. Even more complex approaches in psychotherapy, such as psychoanalysis, have to some degree incorporated such ideas; for example, by focusing on the libido—desire or "wanting" of certain stimuli—or outcomes that motivate and drive decision making in living beings. A second advantage is that focusing on learning mechanisms not only helps to identify basic neurobiological mechanisms and their potential alterations in mental disorders but also allows emphasis of the variety and diversity of human behavior. Indeed, by identifying such basic learning mechanisms and providing a plausible and rational account of how they modify behavior, the resulting explanatory models of human behavior leave ample space for individual experience in the development and maintenance of mental disorders: on the one hand, such disorders can be promoted by basic alterations in learning mechanisms (e.g., an increased focus on negative outcome that can render subjective experiences gloomy), and on the other hand, exactly the same learning mechanisms are directly informed by personal experiences and hence social interactions. Indeed, learning from reward and punishment are two of the research domains suggested by the current research agenda of the National Institutes of Health (Insel et al., 2010).

Moreover, focusing on such basic dimensions has a much longer and more complex history, which includes attempts to map emotions by valence and arousal, resulting in at least two dimensions, one representing positive affect and ranging from highly arousing positive to boringly negative emotions, and the other charting negative affect, which ranges from highly arousing negative to boringly positive emotions (Crawford and Henry, 2004). Of course, philosophers such as Epicurus have long suggested that human beings strive to minimize pain and to maximize positive affect (Ingwood and Gerson, 1994). And the debate on whether human behavior is driven by such basic motivational forces or rather by rational insight has a

long-standing history in philosophy of mind and in philosophical anthropology. We will explore some of these traditions, not because they are inter-
esting by themselves (what they are), but because some of these traditions
can help to caution us not to rely on oversimplified models and because
they may reveal aspects of human behavior that are otherwise out of sight.
Thus, a main aim of this book is to explore how learning mechanisms and
their currently identified neurobiological correlates help us to understand
mental disorders. But before we further explore these topics, we need to
define two concepts just mentioned: what are "mental disorders," and what
do we mean when we talk about "understanding" them?

Let us start with the first topic, which has been the focus of considerable
controversy during the past years. Like any field of research, psychiatry and
its nosology were rather simple when modern scientific classifications were
developed more than 100 years ago (figure 1.1). Basically, at this point there
was a sixfold (three columns × two rows) distinction in place.

In the first column, disorders with a known organic cause were described
and categorized according to whether this cause has an *acute* or a more

Exogenous psychoses (brain organic syndromes)	Endogenous psychoses	Variations
Acute e.g. delirium	The group of schizophrenias	Neuroses (trauma & conflict-related causes)
Chronic e.g. dementia	Major affective disorders (unipolar & bipolar depression)	Personality disorders (traits)

Figure 1.1
Traditional classification of mental disorders.

chronic impact on the brain. Basically, *acute* deliria and other forms of acute organic mental disorders (nowadays often called organic hallucinations, organic paranoia, etc.) were distinguished from more *chronic* forms of mental disorders such as dementia. The former often are associated with clouding of consciousness and disorientation, the latter with impairments of memory function. These disorders were called "organic" because, as in the case of Alzheimer's dementia, visible alterations were observed in brain anatomy (atrophy) and in the structure of cellular and intercellular space (fibrils and tangles). Deliria were called "organic" syndromes not because the exact biological correlates of, for example, delirium tremens were known, but rather because it was plausible that alcohol withdrawal has a direct ("organic") impact on the brain. These two major categories are still reflected in the main International Statistical Classification of Diseases and Related Health Problems, 10th revision (ICD-10), blocks F0 and F1 (World Health Organization, 2011).

In the second column, a fundamental distinction was put in place by Emil Kraepelin, who studied mental disorders that appeared to have a similarly devastating effect on human behavior as observed with the above listed "organic" brain disorders (although in his time, no definite organic cause was found by looking at postmortem brains). He distinguished between disorders that mainly interfere with affect and those that mainly impair cognition. Cognitive impairments were the hallmark of so-called dementia praecox (later renamed schizophrenia), and altered emotions characterized so-called manic-depressive illness (with depression limiting affective responses to the negative range and mania doing the same in the positive range). Here, Kraepelin followed the intuitive distinction between "thinking" and "feeling" and claimed that dementia praecox mainly displays a chronic, continual cause (with Kraepelin largely ignoring the negative effects of prolonged hospitalization of his patients without any treatment), while manic-depressive illness was supposed to follow a cyclic course of illness (Kraepelin, 1913, 1916). We will later focus more on current criticisms of this dichotomy, which is today still reiterated in the distinction between schizophrenia and bipolar disorder in the *Diagnostic and Statistical Manual of Mental Disorders*, 5th edition (DSM-5, published by the American Psychiatric Association (2013), as well as in the major blocks F2 and F3 of the ICD-10 promoted by the World Health Organization (2011).

Finally, in the third column a major group of disorders was listed that were not really regarded as diseases (i.e., major mental disorders), but rather as "reactions" to common stress factors or "varieties" of human behavior. Such disorders included the effects of traumatic experiences as well as distinct personality traits, nowadays classified as a variety of disorders including personality disorders and previously so-called neurotic disorders, which are currently grouped in ICD-10 under the blocks F4 and F6.

However, with all due respect to traditional and current disease classifications, do we really believe that all of these disorders constitute "diseases" in the medical sense? This question points to a discussion about the disease status of mental disorders, which was vivid during the reform of psychiatry in the 1970s and early 1980s. For example, Szasz (1970) criticized the medical disease concept of mental disorders and claimed that in the absence of clearly defined neurobiological correlates, which help to distinguish a mental disorder from a normal, healthy mental state, psychiatry does not describe diseases (with a potential exception of dementia) but instead sanctions social norms and falsely labels socially unwanted or unacceptable behavior as medical problems. Such criticism sparked a long-lasting controversy among medical philosophers, researchers, and clinicians. For the sake of clarification, it is helpful to review the distinction between the concepts of "disease," "illness," and "sickness." The medical aspect of any major malady is commonly called a "disease," the subjective experience contributes to the "illness" experience, and the impairment of social participation is usually called a "sickness" (Sartorius, 2010; Heinz, 2014). We and others have suggested that a clinically relevant mental malady requires the presence of medically relevant symptoms (thus fulfilling the disease criterion) as well as individual harm, which could either be due to a state of suffering from the aforementioned symptoms (the "illness" experience) or a severe limitation of social participation (as in the case of dementia, where a person may not subjectively suffer from memory loss but is unable to perform necessary activities of daily living such as personal hygiene or food intake). We thus conceptualize clinically relevant mental maladies as the combination of disease symptoms with either individual suffering (the illness experience) or impaired social participation (sickness). However, if we do so, a question arises: Which key symptoms of mental disorders can be regarded as medically relevant indicators of a disease and hence as necessary but not sufficient indicators of the presence of a clinically relevant mental malady?

In general medicine, a dysfunction is considered to be a symptom of a disease if it is relevant for individual survival; hence, being unable to roll your tongue, although being a genetically determined dysfunction, is no symptom of a disease, because rolling your tongue is irrelevant for survival. In contrast, being unable to swallow, another dysfunction of the tongue, is highly relevant for human life and is therefore rightly considered to be a symptom of a disease (e.g., a stroke or some impairment of cranial nerves).

Indeed, the medical philosopher Christopher Boorse (1976, 1977, 2012) suggested that in mental disorders, the disease criterion is fulfilled if functions are impaired that are relevant for individual survival. Boorse further suggested that functions relevant for procreation of the species should also be included among functions whose impairment indicates a disease; however, this latter proposal is highly controversial, as it may suggest that subjects who refrain from having children, have a sexual orientation toward their own gender, or have any other reason for decreased rates of procreation suffer from a mental disorder—a point that we and others have sharply criticized. Indeed, we strongly feel that medicine should focus on the health of the individual and abstain from forcing subjects to behave in a certain way in order to fulfill any perceived or constructed need of the "species."

But even if we limit our account of medically relevant functions to those necessary for individual survival, how can such functions be identified? Philosophers usually stop at this point and leave the further elaboration of such definitions to clinical practitioners, who often do not care about the proposals of their philosophical counterparts. Therefore, accounts of basic mental functions and hence of key symptoms indicating a potential mental disorder vary considerably. One way to solve this question, which we feel to be promising in this context, suggests to abandon theoretical construction of potentially relevant mental functions and instead to perform a "pragmatic turn" and to analyze whether key symptoms already used to classify mental disorders (figure 1.2) indeed describe impairments of mental functions generally relevant for human survival in multiple settings and situations. Doing so, it is easy to see that delirium tremens is characterized by clouding of consciousness and disorientation, two functions (consciousness, orientation) that are generally (universally) relevant for human survival. Indeed, it is easy to imagine a multitude of situations (e.g., crossing a road in New York or walking through a jungle in Borneo) in which clouding

1. Vigilance
2. Orientation: person/place/time
3. Understanding communication (including proverbs)
4. Concentration (100 − 7: repeatedly subtract 7 starting from 100)
5. Short-term memory (3 concepts/10 min.)
6. Long-term memory

Figure 1.2
Key symptoms to diagnose acute and chronic brain organic syndromes. Concentration is tested by, e.g., repeatedly subtracting the number 7 from 100.

7. Formal thought disorder (coherence, speed, subjective inhibition)
8. Delusions (delusional mood, delusional perceptions, systematic delusions)
9. Ego disorders (thought insertion, thought broadcasting, thought blockade)
10. Hallucinations (visual, acoustic, commenting voices, voices arguing, commanding voices)
11. Obsessions and compulsions
12. Mood (elevated, depressed, anxious, affective resonance, early morning depression)
13. Drive/motivation (reduced, inhibited)

Figure 1.3
Key symptoms of psychotic and affective disorders (e.g., schizophrenia and bipolar disorder).

of consciousness and disorientation can generally endanger the subject's life (remember that in order to diagnose a clinically relevant mental malady, such symptoms indeed need to cause individual harm, be it that the persons suffers from being disoriented or that she cannot perform activities of daily living such as contacting other persons, etc.).

The situation is more complicated when we look at the key symptoms that to date are used to diagnose schizophrenia or bipolar disorder (figure 1.3). While hallucinations may in general impair the ability of a human being to survive (imagine visual hallucinations that make it very difficult to walk through a room, let alone a city), many hallucinations occur at single time points and carry specific messages that do not directly cause disorientation or a general inability to behave in a given environment (e.g., voices commenting from time to time positively on a subject's performance). However, it can be very difficult to live in a common world with others once, for example, acoustic hallucinations include commanding voices that order a subject to attack someone else. If this subject is indeed following these orders and attacks another person, the attacked subject does not know whether this was a decision of the attacker or whether the attacker was just following the hallucinated orders. Likewise, if during a major depression a person is unable to feel joy (e.g., if her child or grandchild is born) or—in mania—is unable to feel grief even though her best friend has died, it will be very difficult to live in a common world with others. The key symptoms of our major medical disorders thus either directly jeopardize individual survival or (at least) impair basic activities in a world that is shared with

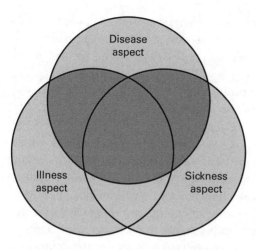

Figure 1.4

Concept of a clinically relevant mental malady. A clinically relevant mental malady (shaded area in this illustration) should only be diagnosed if medically relevant
symptoms are present (the disease aspect of a mental malady, top circle) and either
cause individual suffering (the illness aspect of a mental malady, lower left circle)
or a severe impairment of activities of daily living and hence social participation
(the sickness aspect of a mental malady, lower right circle). Therefore, only disorders
located in the shaded area are diseases defined by medically relevant symptoms that
have harmful consequences for the individual (suffering or severely impaired activities of daily living) and hence constitute clinically relevant mental maladies.

other human beings. However, the latter argument only holds if we assume
that it is a key part of human nature to live with others (as we do). Then,
functional impairments such as imperative voices or the inability to feel joy
or grief can be counted as major dysfunctions, which constitute the medical aspect of a disease.

At this point, we would like to emphasize that the manifestation of a
symptom of a disease, for example a hallucination, is not enough to diagnose a clinically relevant mental malady—the person also has to suffer
from it or the hallucination has to impair basic activities of daily living (figure 1.4). One of our patients entered the hospital by stating that we should
"leave his voices alone": "I speculate at the stock exchange, the voices tell
me where to invest and so far their recommendations have always been
useful!" In such cases (and in the absence of further symptoms, personal
suffering, and an impairment of daily living), a doctor may well diagnose
a medically relevant dysfunction (voices are acoustic hallucinations and

hence a disturbance of generally life-relevant perception), but—as we feel—
the same doctor should abstain from diagnosing a clinically relevant mental malady, because the person reporting the hallucinations neither suffers
from them nor is impaired in coping with her activities of daily living.

The resulting concept of a clinically relevant mental malady is, however,
much narrower than all the behavior patterns classified in DSM-5 or ICD-10
as mental disorders. For example, it may be hard to argue why experiencing
social phobia only when speaking in front of a major audience should be
counted as an impairment of a medically relevant function that is generally
relevant for human survival (or even for living with others in a common
world)—there are, as we feel, many human subjects who are not able to give
speeches in front of other people who are nevertheless able to live a satisfy-
ing life. On the contrary, this example shows that the empirical application of philosophical concepts always results in descriptions and definitions
with "vague" boundaries: what at a certain point in time did not represent
a mental function necessary for survival (or for basic interactions with others) may become so in the future (or even in contemporary society, given
the abundance of PowerPoint presentations required to present yourself and
your work in front of others). What we want to emphasize here is that mental disorders (such as social phobia manifesting itself when speaking in front
of an audience) are a much broader category that includes multiple states of
human suffering, which is why medical diseases (such as delirium tremens
or bipolar disorder) are only a subgroup of such disorders characterized by
their universally high relevance for human life. We indeed feel that medically relevant diseases should only be diagnosed if the functions impaired
in such states (as indicated by key symptoms to diagnose the disease) are
relevant for the individual person's survival or her basic ability to interact
with others. It is open to public debate which exact functions should be
classified as medically relevant, and medical doctors as well as psycholo-
gists have professional knowledge that helps to articulate such suggestions;
however, the final decision is up to an open dialogue in society, which
needs to include the views of patients and their relatives, lawmakers, and
the broader public. In the current volume, we will focus on major mental
disorders (i.e., such states that have traditionally been and in our view are
rightfully called diseases: addiction, psychoses, and major affective disorders). However, while doing so we will touch on related phenomena such as
anxieties and other negative mood states, cravings, obsessions, and experiences of alienation.

Finally, we need to discuss what we mean when we talk of "understanding mental disorders." As suggested earlier, modern computational tools

help us to analyze behavior on a much more complex level than previously possible. For example, such studies can use rather straightforward decision-making paradigms, in which you can press a left or right button and are rewarded probabilistically (i.e., in only 80% of all times once you press the currently better button), and after a while, unannounced switches will occur after which you have to learn the better choice again. As simple as this task may seem to be, it can be solved with a multitude of strategies: you can compute the feedback immediately after your choice (e.g., if you press the left-hand button you memorize the reward or punishment that follows your choice); alternatively, you can also compute the reward you *would have* gained if you *had* chosen the other option (the right button, which you did not press). Moreover, you can react more strongly to punishments (unexpected losses) than to rewards, which is a behavioral pattern that many individuals display. Finally, you can start counting in order to find out how many negative feedbacks have to occur to make it very likely that another unannounced change has occurred and you better switch your choices (in the above given example, it is now better to press the right-hand button, even though this choice was previously punished most of the time). On the basis of individual behavior patterns, researchers can build different mathematical models and find out which one of them can best explain the respective choices of the individual proband (more about that in later chapters). Understanding behavior with such computational tools hence means that you can analyze and model it. Furthermore, it suggests that human decision making may generally follow such mathematically describable patterns, which can be simulated by a computer and hence explained in mathematical terms.

However, understanding mental disorders traditionally had a much wider meaning: Dilthey (1924) suggested distinguishing between an *explanation* of behavior (e.g., in the above given example by mathematical models or, in his time, by assumptions about organic causes of behavior) and *understanding* human behavior, which requires some degree of empathy and relies on your personal experiences. To support his view, Dilthey suggested that if you want to learn something about human psychology, reading novels will have a much more profound effect compared to reading the results of experiments provided by the then-developing scientific psychology. What was true for the end of the nineteenth century to some degree still appears to be true in our contemporary times: the "how it is to be" aspect of any human behavior can easily get lost in mathematically articulated "computational" descriptions of human behavior. However, focusing on learning mechanisms and their modification of and by experience may, as we feel, help to bridge the gap between a mathematical *explanation* and a subjective *understanding* of human behavior and its alteration in mental disorders: all people know how it "feels" to learn from experience, all have experienced rewards and punishments, received gains and suffered from losses, anticipated positive and negative outcomes, and went through hope and despair. As we will try to show in this volume, applying modern computational techniques to explain behavior does not mean to ignore subjective experience but rather to respect its uniqueness by trying to better understand how individual human experience is modified by learning mechanisms throughout the life span. Also, we want to emphasize that in spite of using computational techniques to explain human behavior, a human being, its body and its central nervous system, are much more complex than any computer simulation will ever reveal. Even if we focus on the central nervous system and its main organ, the brain, this living organ is in a dynamic interaction with a multitude of somatic inputs (e.g., from sensory organs, hormonal systems, etc.), which reflect states of the human body and its environment and represent extremely complex information in digital as well as analogous ways. If we want to identify computations performed by a brain by using in vivo imaging techniques and by simulating the decisive steps with our most advanced mathematical techniques, we look at just a minor part of all the rhythms and dynamics occurring in this organ. This remark should protect us from any kind of reductionism: science per se has to reduce the complexity of any given situation in order to find regularities that at best can be described using some mathematical formula or other ways to predict future events. We thus form a qualitative or quantitative (mathematical) model of our environment, which for the sake of parsimony is less complex than the facts in focus. However, trying to explain human behavior by using mathematical models and linking them with brain signals to help understand human experiences can reduce the complexity of these experiences without ignoring all that cannot be captured by such models: multifold and extremely diverse, valuable and highly important parts of human life uncharted by scientifically unavoidable, pragmatic reductions of complexity via experimental settings and paradigms.

The current volume will proceed in three parts. First, it will explain Pavlovian and operant conditioning, Pavlovian-to-instrumental transfer, and the distinction between habitual and goal-directed decision making (chapters 2–4). This part will help to illustrate and explain basic learning mechanisms and their respective effects on human behavior. Following a dimensional approach in psychiatry, our account will not focus on single disorders but rather explain basic mechanisms that are supposed to be at work in a multitude of mental states and disorders. In the second part (chapters 5–9), we will explore how far such basic dimensions of mental disorders carry us when we try to explain key syndromes of mental disorders, such as craving and loss of control in addiction, positive and negative mood states in affective disorders, as well as experiences of being alienated with respect to one's own embodied self and the environment in psychotic states. Again, we will not focus on single distinct disease categories (schizophrenia vs. brief psychosis vs. schizoaffective disorders vs. psychotic experiences in organic hallucinations) but rather explain key mechanisms with respect to their syndromatic correlations. At the end of chapters 7–9 and in the last part of the current volume (chapter 10), we will explore clinical and therapeutic implications of such accounts. Computational approaches in psychiatry that focus on basic dimensions of human decision making, specifically on learning mechanisms, may thus offer a way to look at human diversity, emphasize personal experiences, and help to explain why not one mental disorder is effectively like another.

2 Basic Learning Mechanisms: Pavlovian Conditioning in Mental Disorders

"When it gets dark, the sky turns gray and I walk alone in the streets, and when I then pass by a bar, see the warm yellow light and hear the clinking of the glasses, I am lost." This is how one of our alcohol-dependent patients described situations in which he relapsed in spite of his conscious decisions to remain abstinent. The warm yellow bar light and the clinking of the glasses can be understood as Pavlovian conditioned cues that elicit an urge (a so-called craving) to consume the drug of abuse, in this case alcohol. Indeed, in his famous experiments with dogs, Pavlov observed that not only food itself but also stimuli that are regularly associated with the presentation of food, such as his own steps when he approached his dogs, can elicit the same response, the production of saliva (Pavlov, 1928). Food is a so-called natural reinforcer, which means that all animals look for food, orient their behavior toward the acquisition of this resource, and if successful tend to repeat the actions that gained this reward. Food thus reinforces behavior aimed at its acquisition. However, during Pavlovian conditioning, the food reward is not actively approached: Pavlov's dogs did not perform specific activities to acquire food, instead, passive encounter with this natural reinforcer elicited responses such as the production of saliva, a natural process that helps to prepare for and facilitate food consumption. Such *unconditioned responses* are not actions performed by the animal in order to acquire food; rather, they are inborn mechanisms that manifest automatically. Neither do animals have to learn to crave for food. Instead, food acts as an *unconditioned stimulus* evoking a hardwired response, saliva production (the *unconditioned response*). Regular pairing of environmental stimuli with the presentation of food, be it the steps of Pavlov himself approaching his dog or the use of a bell (as an artificially introduced signal that announces the availability of food), transform such cues into *conditioned stimuli* that evoke a *conditioned response*—the "hardwired" production of saliva, which is now called a conditioned response simply because it was evoked by a conditioned stimulus and not by the natural reinforcer, food.

When we look at what our patient described, similarities and differences are obvious: beer, the favorite drink of our patient, can be regarded as a natural reinforcer. It is a nutritious drink that includes alcohol, a drug of abuse acting on a variety of neurotransmitter systems, some of them—as will be discussed later—reinforcing alcohol intake. The clinking of glasses, in contrast, is a culturally embedded ritual that is absolutely arbitrary. While the smell of beer is an unconditioned stimulus, which can evoke both saliva production and craving for this drink, the clinking of glasses has certainly nothing to do with alcohol intake before a person learns that it is exactly this ritual that precedes alcohol consumption in many cultures. Here, we thus have a classical conditioned cue (the clinking sound) that evokes a conditioned response, craving for the drink, just as the smell of a preferred beverage acting as an unconditioned stimulus can promote an urge to consume this drink (the unconditioned response). However, our example shows how complex the situation really is. Contextual cues such as the fading of colors during dusk and the feeling of loneliness of our patient appear to play a role. They set a certain mood in which our patient may want to forget about his problems, be with others and consume a drug that improves his negative mood state, soothes him, and helps him to relax.

Moreover, hearing the clinking of glasses, seeing a warm yellow bar light, and wanting a drug is not the same as actually entering the bar and consuming the drug of abuse. Our patient had decided to remain abstinent and, as far as he was able to reflect upon his own intentions, did not consciously question this decision. Rather, he felt driven "against his own better insight" to enter the bar and to order a drink. The confrontation with drug-associated conditioned cues, as it appears, can thus help to explain why and how a certain desire is evoked; however, their motivational effects do not suffice to describe the whole complex series of actions and thoughts that lead to a relapse. We will later discuss how Pavlovian conditioned cues can influence the instrumental actions of a person (e.g., those that are required to enter a certain location, order a drink, pay for it, and consume it). Such effects of Pavlovian cues on instrumental behavior are called Pavlovian-to-instrumental transfer (PIT). Yet before we can address such interactions between Pavlovian stimuli and instrumental behavior, we need to look in more detail at Pavlovian cues and their effects in mental disorders.

Like alcohol craving, panic attacks can also be triggered by Pavlovian conditioned cues. For example, being in the middle of a theater during a performance that you cannot leave without upsetting a lot of people can trigger a panic attack, as it can be elicited when you have to cross a wide-open space or enter a narrow elevator. One of our patients described that her first panic attack manifested when she accompanied an elderly lady, whom she regularly helped to cope with her everyday duties, to a bank. They were standing in line and our patient felt quite uncomfortable both about having to wait when she had to get her own duties done and about her inability to communicate to the elderly lady that she has other things to do and actually needed to leave. The latter thought was not present to her during the situation but only upon later reflection. In the situation itself, she suddenly felt that she could not breathe anymore and that her heart was beating wildly. The feeling was so strong and alarming that she thought she would suffer from a heart attack and she never entered this bank again, being afraid to develop a panic attack again. Patients indeed quickly learn to avoid all situations that can trigger such anxiety-provoking physical mani-festations. Cognitive behavior therapy aims at overcoming such avoidance behavior by exposing the patient to the situation, in which the anxiety will quickly rise and then slowly fade away—but only if the patient manages to stay in the situation instead of avoiding it. Thus, the patient has a chance to "unlearn" the fear response. In Pavlovian terms, the strong rise in alarm-ing bodily sensations is an unconditioned response, while having to wait in line is a conditioned stimulus that gets immediately associated with the very unpleasant feelings of suffocation and fear. The subject tends to avoid both the fear and the contextual stimuli that can act as conditioned cues and trigger the panic response.

With respect to neurobiological correlates of Pavlovian conditioning pro-cesses, most studies focused on such aversive outcomes and observed that aversive conditioned stimuli activate the anterior cingulate cortex, bilateral insulae, and the amygdala, with functional brain activation of the amyg-dala significantly declining over time (Büchel and Dolan, 2000). Further-more, the timing of the presentation of the conditioned cue with respect to the unconditioned stimulus is relevant. In so-called delay conditioning, the unconditioned cue is usually presented at the offset of the conditioned stimulus, and there is no gap between the offset of the conditioned and the onset of the unconditioned cue. In so-called trace conditioning, there is a time interval between the offset of the conditioned cue and the onset of the unconditioned stimulus. Hence, a memory trace has to be produced to "bridge the gap" between the conditioned and the unconditioned stimu-lus, which is reflected in additional activation of the hippocampus in trace conditioning but not delay conditioning (Büchel and Dolan, 2000). Fur-thermore, different nuclei of the amygdala appear to play specific roles. For example, lesions of the central but not basolateral nucleus of the amygdala impair appetitive Pavlovian conditioning (Parkinson, Robbins, and Everitt, 2000).

However, correlation is not causation, and the question remains why a conditioned response is strong enough to clinically manifest as a panic attack. Why did our patient not just experience a rise in blood pressure and heart rate when becoming annoyed while waiting in line, instead of being overwhelmed by the fear of a heart attack and suffocation? Many people accompany others in situations in which they do not feel well, they have to wait in line for hours or feel too shy or too polite to express their dissatisfaction with a situation in which they feel that the person whom they help is not respecting their limited time resources. Usually, none of these subjects ever experiences anything close to panic. Moreover, not only patients but often also medical doctors feel tempted to perform and repeat complex series of somatic examinations to exclude the possibility that a serious somatic disorder causes the feelings of suffocation or heart pain. Excluding somatic disorders carefully (but only once) is indeed useful, because in some rare cases, such somatic problems exist and explain the aversive condition. However, in most patients, no such somatic causes can be found. So what happens? Is there unusual attention focused on common somatic phenomena such as a rise in heart frequency when one feels angry or is there some primary alteration in the physiology of sensory perception in those persons that suffer from panic attacks? In other words, is the conditioned evocation of panic attacks by environmental stimuli simply a case of Pavlovian conditioning or is there a much more complex process at work?

Such questions are hard to answer in retrospect (e.g., by questioning patients with panic attacks about what exactly they felt and when). Prospective studies are helpful, but only rarely possible. One example for such a prospective approach is the study by Godemann and colleagues (2006). It examined patients with vestibular neuritis, a disorder of the inner ear that is accompanied by dizziness and nausea. A high number of patients with vestibular neuritis develop panic and somatoform disorders; however, by far not all patients do so. In this patient group, it was possible to examine what predicts the development of panic attacks: is it the strength of the primary physiologic problem (i.e., the strength of vertigo, nausea, and dizziness) or the subjective fear associated with such symptoms and the ensuing attention focused on these signs? Godemann and colleagues observed that it was not the degree of nausea or vertigo that predicts the development of panic attacks. Neither was it predicted by the degree of fear arising on the first day of an acute vestibular neuritis. However, persistent fear of vertigo or vomiting as well as panic-related thoughts focusing on concerns about one's own health predicted up to 60% of the variance in the development of panic or somatoform disorders. Furthermore, general risk factors such as poor psychosocial integration or a lack of social support contributed to the risk of developing panic disorders (Godemann et al., 2009).

These observations illustrate two important points: first, it is fear and attention directed at somatic symptoms rather than the primary severity of somatic symptoms itself that predicts the development of panic disorders. Second, panic attacks are not simply an unconditioned somatic response that triggers Pavlovian conditioning of arbitrary environmental cues, which then become Pavlovian conditioned stimuli that elicit panic as a conditioned response. Rather, unconditioned responses including nausea or dizziness after vestibular neuritis evoke a complex interaction between these physiologic reactions, specific environmental cues, complex contextual situations and the associated emotions and cognitions. In case such interactions result in a focusing of attention on certain somatic experiences (e.g., nausea and dizziness after dysfunction of the inner ear), excessive salience can be attributed to the somatic symptoms, and panic attacks manifest. The concept of Pavlovian conditioning can then help to explain why further panic attacks are triggered by conditioned cues; however, the development of panic attacks already includes certain cognitions and anxieties, a focusing of attention, and a worry about one's own health.

Another example helps to illustrate how cognitions can turn a common object into a conditioned cue that triggers obsessive thoughts and compulsive actions. One of our patients, a nun, had developed a compulsion of obsessively washing her hands. She was in her late fifties and did not suffer from obsessive-compulsive disorder before her mid-forties. When asked how her problems started, she explained that she was ordered to take care of the towels of all nurses. Whenever she touched a towel in a bathroom, she afterward washed her hands, "because you know where my sisters had her hands before they touched the towels." Here, the towels appear to have become conditioned stimuli that trigger thoughts about hands touching private body parts. These thoughts manifest as obsessions; that is, our patient did not want to think about such issues but found herself to be helpless when trying to stop these unwanted cognitions and emotions. She obsessively thought about how the towels are soiled by hands being in places she did not want to speak about, and she incessantly washed her hands in order to clean herself from the imagined pollution occurring when she touched the towels. In obsessive-compulsive disorders, compulsions are often performed in order to keep obsessive, unwanted thoughts at bay. However, performing the actions is not successful to stop the worrisome thoughts and thus has to be repeated again and again whenever the unwanted obsession reappears. Cognitive behavior therapy aims at exposing the subject to exactly these thoughts, while ensuring that the compulsive acts are not performed. Again, Pavlovian conditioning only explains a certain part of the behavior: towels are not natural objects associated with the anus or genitals, hence associating them with these body parts is an unintentionally acquired process, which turns them into conditioned stimuli that evoke a conditioned response—obsessive thoughts about impurity and a strong desire to clean oneself. Pavlovian conditioning can thus help to explain why towels become such important objects in the life of our patient, while such conditioning processes do not per se help to explain why she worried so much about hands touching towels and why she was not satisfied with carefully washing her hands once but instead repeated the procedure again and again.

The same is obviously true with respect to cue-induced triggering of relapse in drug addiction. Cues can trigger drug craving and simple, hardwired responses such as the production of saliva; however, in order to understand their role in relapse behavior, we have to look at a wider picture. We have to consider specific and general environmental factors and their effect on ongoing activities in order to understand how Pavlovian conditioned cues may trigger relapse. To do so, we have to turn to Pavlovian effects on instrumental behavior.

Pavlovian-to-Instrumental Transfer

Whenever we talk about actions performed by an individual to reach a certain goal, we speak about instrumental behavior (i.e., the action is supposed to be instrumental to attain a certain reward or avoid an unpleasant outcome). Operant conditioning describes how rewards and punishments guide our behavior. Basically, behavior will occur more often when it is rewarded and less often when it is punished. Being afraid to be punished but successfully avoiding punishment will also reinforce actions that helped to avoid the unpleasant outcome. Positive reinforcement therefore consists of gaining a reward, while successfully avoiding punishment reinforces the respective actions that help to evade punishment and is hence called negative reinforcement (i.e., it results from avoiding punishment).

In the next chapter, we will explain in more detail which brain regions and neurotransmitter processes are associated with instrumental behavior. At this point, we will focus on the effects of Pavlovian conditioned cues on instrumental behavior, the so-called Pavlovian-to-instrumental transfer

(PIT) effect. In the clinical examples given above, we have already seen that Pavlovian conditioning usually occurs within a wider framework of behavior. Generally, human behavior can be understood by the S-O-R-C scheme (Kanfer and Saslow, 1965): a stimulus (S), for example a food cue, triggers—depending on the state of the organism (O), for example being hungry or not—a certain response (R), which can be an inborn response such as saliva production or an acquired goal-directed action, which is then followed by a positive or negative consequence (C). Indeed, intentional acts are usually triggered by specific environmental stimuli or more general aspects of the social context in which an individual is situated.

A subclass of such stimuli is called Pavlovian conditioned cues because of their ability to elicit a conditioned response (such as the production of saliva in a hungry dog), which is usually inborn and hence hardwired in the central nervous system. However, Pavlovian conditioned stimuli can also impact on ongoing instrumental behavior, even if the instrumental behavior was acquired independently of Pavlovian conditioning. For example, a person can be instructed to collect certain kinds of objects (e.g., shells of a certain shape and color) and to reject others (e.g., shells of a different color or shape). Whenever correct shells are collected, the individual will be reinforced; whenever a false shell is taken, the individual will be punished (e.g., by losing money). The person thus performs instrumental actions in order to gain as much money as possible. This instrumental behavior can be influenced by presenting unrelated Pavlovian cues in the background. For example, when shells are presented on the screen, the background of the screen can either be neutral or present a Pavlovian conditioned stimulus, for example a fractal (i.e., a complex colorful picture that acts as a Pavlovian conditioned stimulus). For instance, a fractal with a certain color may have previously been paired with passively gaining money (i.e., whenever this fractal appeared on the screen, the person was informed that she automatically won 2 euros or dollars) (figure 2.1). A fractal with a different color can have been paired with no outcome, and a third fractal with yet another color was regularly associated with passively losing money. Passively winning or losing money after the presentation of these fractals is thus completely independent of actions associated with collecting shells in order to gain a reward. By passively pairing fractals of different colors with a positive or negative monetary outcome, these fractals become positive, neutral, and negative Pavlovian stimuli. Notably, pairing these Pavlovian cues with the unrelated instrumental task will bias subjects in their decision making. A similar phenomenon can be observed in large shopping malls: usually, pleasant music is played in the background in order to induce a

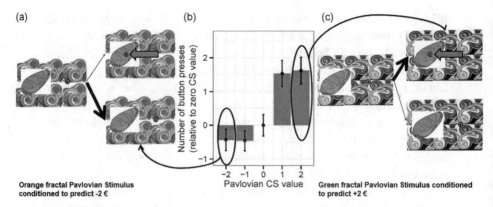

(a)   (b)   (c)

Orange fractal Pavlovian Stimulus   Pavlovian CS value   Green fractal Pavlovian Stimulus conditioned
conditioned to predict -2 €   to predict +2 €

Figure 2.1

Pavlovian-to-instrumental transfer. In the instrumental task, one has to learn which shells to collect and which ones to reject in order to gain as much money as possible. Absolutely unrelated to this instrumental task, colorful fractals are presented in the back of the screen, which have been passively paired either with monetary loss (left side, panel a) or no change or monetary gain (right side, panel c). These fractals thus act as independent, Pavlovian conditioned background stimuli (CS) and interact with instrumental behavior in the unrelated task to select the correct shells (center, panel b) (Garbusow et al., 2014). (a) The proband has learned to collect (i.e., approach) a specific shell by repeatedly pressing a button that moves the cursor onto the respective shell. The unrelated Pavlovian stimulus (the fractal) presented in the background is aversive, because it has previously been paired with passive monetary loss (i.e., whenever the brown fractal appeared during Pavlovian conditioning, the proband was unable to react and was informed that he just lost 2 euros). (b) The number of button presses (instrumental response) as a function of the value of the respective Pavlovian background stimulus (–2 €, –1 €, 0 €, +1 €, +2 €). A Pavlovian background cue associated with "0 €" is the neutral condition (i.e., presentation of this fractal was not previously associated with either passive monetary gain or loss). With increasing value of the Pavlovian stimulus in the background, the number of button presses increases, while decreasing values of the Pavlovian background cue are associated with a reduced number of button presses in the unrelated instrumental task. Note that the probands were instructed to focus on the shells and not the background stimuli. (c) Combining the shell with a positive (i.e., previously rewarded) Pavlovian cue (i.e., the proband always passively gained 2 euros whenever this blue fractal appeared on the screen) in the background of the screen increases approach behavior in the absolutely unrelated instrumental task, as shown in (b) by the increase in the number of button presses to acquire the respective shell. Source: Modified according to Garbusow et al. (2016a).

positive mood in the potential consumers, who are then supposed to be more inclined toward spending money. Here, independent background stimuli promote approach behavior.

This mechanism can be illustrated by a further example: Imagine walking through a forest on a bright spring morning. The birds are singing, the sun is shining through the leaves, and when you hear a noise on the forest floor, you may feel inclined to see whether it is a rabbit or some other animal moving through the forest. Now imagine the same scene during a cold November night. It is dark, cold, and drizzling, you cannot see very far, and hearing a noise behind you may make you startle and wish to run away rather than motivate you to explore the source of this noise. Contextual cues thus influence your behavior; in this example, affectively negative stimuli promote a tendency to withdraw and inhibited the tendency to approach and explore a situation. Notably, the same effect can be observed when combining Pavlovian conditioned cues with instrumental behavior: when the above described fractals, which have previously been associated with monetary reward, are presented in the background of a screen, a proband will show more effort (a higher number of button presses) tending to collect the shell presented in the independent instrumental task, even if collecting this specific shell is punished rather than rewarded (Garbusow et al., 2014, 2016a). Likewise, presenting a Pavlovian conditioned fractal that has previously been associated with monetary loss will decrease the effort for the instrumental behavior (or increase the tendency of a subject to withdraw, i.e., not collect a shell), even if collecting this shell would lead to monetary gain (Huys et al., 2011).

Pavlovian cues thus interact with apparently independent instrumental behavior, and the resulting PIT effect can help to explain why the clinking of glasses motivates our earlier patient to enter the bar, order a drink, and consume the alcoholic beverage. The Pavlovian conditioned cues can stimulate a conditioned reaction, for example the craving to consume alcohol; however, this motivation alone does not suffice to explain why our patient indeed entered the bar—consciously, he wanted to remain abstinent. To explain the effects of the appetitive Pavlovian cue (the clinking sound) on complex approach behavior, one has to refer to PIT effects as described above. The clinking of glasses heard by the patient who consciously decided to remain abstinent will bias his behavior toward approaching the alcohol reward, and he feels driven or even "compelled" to enter the bar rather than avoiding it. Pavlovian conditioned cues can thus have a decisive influence on the actions of a person who is ambivalent whether to follow her urges or conscious decisions. Indeed, such effects of Pavlovian background cues on instrumental behavior are particularly strong when reward outcomes are uncertain (Cartoni et al., 2015).

With respect to neurobiological correlates, we have to distinguish between two forms of PIT: in the outcome-specific form of PIT, the Pavlovian cue has been conditioned with the same rewarding outcome that can also be gained when performing the instrumental response (e.g., the smell of wine promotes ordering and consuming a glass of wine but not a glass of lemonade), while in the general form of PIT, the Pavlovian cue has been conditioned to a positive outcome that is not associated with the outcome available by the instrumental action (e.g., upbeat music played in a shopping mall motivates customers to buy certain goods). General PIT thus appears to promote instrumental actions by generally modulating arousal, while outcome-specific PIT may facilitate the retrieval of particular actions on the basis of their outcome (Corbit, Janak, and Balleine, 2007). Animal experiments and human studies suggest that activation of the basolateral amygdala and the ventrolateral putamen contributes to outcome-specific forms of PIT, while the central nucleus of the amygdala is implicated in general PIT (Corbit and Balleine, 2005; Prevost et al., 2012). Impairment of ascending dopaminergic projections due to inactivation of the ventral tegmental area (VTA) attenuated both general and specific forms of PIT (Corbit et al., 2007).

There appear to be interindividual differences in the degree to which a subject is influenced by Pavlovian background cues. For example, detoxified alcohol-dependent patients were generally more driven by PIT effects than age- and gender-matched healthy control subjects (Garbusow et al., 2014, 2016a). We will discuss later whether such differences in individual vulnerability toward specific learning effects are most likely heritable or acquired, for instance due to the effects of drugs of abuse on basic learning mechanisms. Here, it will suffice to emphasize that such individual PIT effects can be quantified; for example, in our instrumental task, a person has to repeatedly press a button to acquire a certain shell, and the number of button presses can be counted. A strong PIT effect is characterized by (1) a high number of button presses performed by the individual in an independent instrumental task when being confronted with positively conditioned Pavlovian cues in the background, and (2) by a strong reduction in the number of button presses when being exposed to negatively conditioned Pavlovian background stimuli. A weak PIT effect, in contrast, is characterized by a small change in instrumental behavior (here the number of button presses when confronted with positive, neutral, or negative Pavlovian conditioned cues). Indeed, the strength of this quantifiable PIT

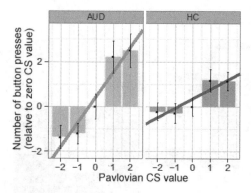

Figure 2.2
Differences in the strength of PIT effects between alcohol-dependent patients (AUD, alcohol use disorder) and healthy controls (HC). The strength of the PIT effect is quantified by the effect of positive (i.e., previously paired with passively winning 1 or 2 euros) versus neutral and negative (i.e., previously associated with losing 1 or 2 euros) Pavlovian background cues (CS, conditioned stimulus) on the unrelated instrumental task, which requires the subject to repeatedly press a button to collect a rewarding shell (or to abstain from pressing the button to avoid an unfavorable shell). Source: Modified according to Garbusow et al. (2016a).

effect was significantly stronger in detoxified alcohol-dependent patients compared to healthy control subjects (Garbusow et al., 2014), suggesting that alcohol-dependent patients are particularly vulnerable to positive and aversive background cues (figure 2.2).

This observation may help to explain why our patient, who was walking alone through the streets and felt lonely at the beginning of the night, was strongly drawn toward entering a bar and ordering a drink when confronted with Pavlovian conditioned stimuli associated with alcohol intake.

One way to deal with such PIT effects is training to overcome the approach bias elicited by alcohol cues. In fact, Wiers et al. (2011) observed that presenting alcohol versus neutral cues is associated with a tendency to approach such cues in alcohol-dependent patients. Probands were told to use a joystick in order to push pictures away when presented in landscape format and to pull them toward themselves when the pictures were shown in portrait format (figure 2.3).

The authors observed that alcohol-dependent patients had higher reaction times when required to push alcohol pictures away, while it was easier for them (i.e., they were faster) when they had to pull the alcohol pictures toward themselves (Wiers et al., 2011). This alcohol approach bias was

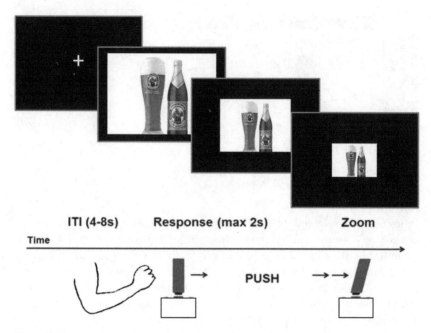

Figure 2.3
Zooming joystick task. Whenever a picture is presented in landscape format, the proband has to push the joystick away from himself (here represented by a movement of
the joystick to the right), and as a consequence, the picture will become smaller and
smaller. If the same beer bottle (or any neutral picture) appears in portrait format,
the proband has to pull the joystick toward himself, and the picture will be enlarged
(ITI, inter trial interval).

overcome by specific training: in the verum condition, alcohol pictures were
presented 90% of all times in the landscape format, which requests probands
to actively push the picture away and thus teaches them to automatically reject alcohol cues, while in the placebo condition, alcohol pictures
were presented as often in landscape as in portrait format, and hence probands pushed them as often away as they pulled them toward themselves
(figure 2.4).

Notably, only six training sessions appear to be enough in order to substantially reduce relapse rates within the next year (Eberl et al., 2014). The
therapeutic success was not simply explained by reaction-time changes,
suggesting that the training may not just alter an alcohol approach bias but
potentially affect other mechanisms relevant for relapse including alterations in the value of alcohol cues and PIT effects.

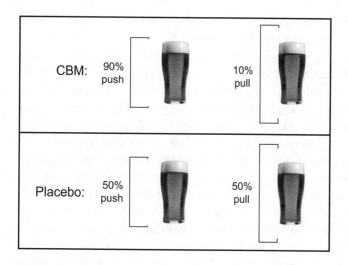

Figure 2.4
Design of the zooming joystick task. In the verum condition aimed to modify the approach bias elicited by alcohol cues (CBM, cognitive bias modification), alcohol pictures have to be pushed away in 90% of all trials in accordance with the format in which they are presented, while in the placebo condition, the same cues have to be pulled toward the subject as often as they have to be pushed away.

Neurobiologically, the Pavlovian conditioned alcohol cues activated the ventral striatum, a core area of the so-called brain reward system, the medial prefrontal cortex and the amygdala (Wiers et al., 2014). The amygdala has been implicated in conditioning processes, not only with respect to aversive but also with respect to positive Pavlovian stimuli. Notably, reduced activation of the amygdala was found in patients exposed to verum but not placebo training. This finding suggests that habitual approach tendencies elicited by Pavlovian conditioned cues can be overcome by training the automatic rejection of such stimuli. Combining alcohol pictures with an instrumental task, which requires pushing such cues away, may thus help to overcome the established association between the Pavlovian cue (the alcohol picture) and the Pavlovian conditioned response (the approach motivation) and its effect on instrumental behavior (the approach bias), and instead establish a new bond between environmental stimuli (the alcohol pictures) and instrumental action (the rejection of or pushing away of the pictures).

In summary, Pavlovian learning mechanisms help to explain how previously neutral environmental stimuli can become Pavlovian conditioned cues that evoke conditioned responses; for example, an urge to drink when hearing glasses clink, disgust when having to touch a towel in a bathroom, or fear of a heart attack when entering a certain place. Because of the complex nature of human behavior, the explanatory power with respect to the development of mental disorders is limited; however, Pavlovian-to-instrumental transfer effects appear to play a prominent role in relapse behavior among addicted patients. Furthermore, understanding such rather simple Pavlovian-to-instrumental transfer effects and how to deal with them can inspire new treatment methods, which go beyond simple cue-exposure and involve active rejection of drug cues or other rather automatically performed activities aimed at the promotion of a healthy lifestyle. Brain areas implicated in Pavlovian conditioning processes include the ventral striatum, a brain area associated with processing of salient, important stimuli indicating potential reward and punishment; the medial prefrontal cortex and cingulate cortex, which are associated with error monitoring and the interaction between emotion and cognition; and the amygdala, a so-called limbic brain area that is involved in conditioning processes and the experience of emotions. Serotonergic modulation of the amygdala has often been associated with the experience of anxiety and fear; however, other neurotransmitter systems such as dopamine, of which we will learn much more soon, have also been implicated in processing aversive stimuli in this brain area (Kienast et al., 2008). Indeed, anxiety and other negative emotions are not simply associated with activation of the amygdala by aversive stimuli; rather, the interaction between the amygdala and brain areas associated with emotion regulation such as the cingulate cortex correlated with the degree to which unpleasant stimuli elicit negative mood states (Kienast et al., 2008).

These findings help to illustrate two points: First, clinically relevant psychological phenomena such as anxiety, fear, or other negative mood states have complex neurobiological correlates. The amygdala is more than a "fear center," likewise, the ventral striatum does not represent "the reward system"; rather, both brain areas are embedded in complex neurocircuits that include cortical and subcortical brain areas and contribute in varying degrees to human cognitions and emotions. Second, as complex as the underlying neurobiology may be, it is possible to identify key brain areas implicated in basic learning mechanisms such as Pavlovian conditioning. We will hear more about dopamine, serotonin, and other neurotransmitter systems interacting with such brain areas in the next chapter, when we discuss goal-directed versus habitual aspects of decision making based on reward and punishment.

3   Reward-Dependent Learning

Operant conditioning appears to have rather simple effects: rewarded actions are repeated, while any activity that is followed by punishment will not be carried out again. Drug intake, being subjectively rewarding, should be repeated; however, when it leads to long-term negative consequences, it should be avoided as soon as these aversive consequences manifest. Or so it should be. Given that addicted subjects do not cease to consume drugs that have aversive long-term consequences, a series of theories tries to explain this behavior. One theory suggests that addicted subjects are too "shortsighted" and prefer short-term rewards over long-term benefits, thus focusing on the immediate effects of drug consumption instead of orienting their behavior toward long-term positive consequences of abstinence. This behavior pattern is often described as "delay discounting" or "impulsivity," although, as we want to warn, impulsivity is a vague construct that besides delay discounting includes rather unrelated aspects such as being unable to stop ongoing behavior or reacting aggressively when provoked (Heinz et al., 2011). An alternative theory suggests that drugs of abuse "highjack" the reward system, thus rendering drug-related rewards much more salient or strong compared to non-drug-associated reinforcement (Goldstein et al., 2010). Before we address such questions in more detail, it is worth looking at basic theories of reinforcement-related learning. We will address several points.

First, is reward-dependent learning really dependent on rewards? This question may sound like nonsense; however, it touches upon an important problem, which arises when animal experiments are translated into human research. In fact, behaviorists did not speak of reward but rather of reinforcement: any outcome that increases the probability of repetition of an action is accordingly called a positive reinforcer, while any outcome that reduces the occurrence of such behaviors is a punishment. If an aversive condition is abolished by a certain activity, a so-called negative reinforcement takes place, which consists of the abolishment of the unpleasant state and increases the likelihood that such an activity is displayed again upon confrontation by the aversive situation (Skinner, 1953).

When dealing with animals, experimenters are usually cautious and avoid attributing subjective emotions such as pleasure to the animals. This is usually not the case when the same learning models are applied to human beings. A straightforward reward learning theory indeed suggested that all drugs of abuse release dopamine and that dopamine release is inherently pleasant, thus reinforcing drug consumption, while blockade of dopaminergic neurotransmission interferes with hedonic pleasure and is hence aversive (Wise, 1988). This model was applied to explain drug addiction, because drugs of abuse tend to release much more dopamine than any natural reinforcer (such as food or sex) and are thus supposed to affect behavior more strongly than other outcomes. Accordingly, animals as well as human beings who consume drugs would pursue drug consumption and prefer it to other outcomes such as food or sexuality, because positive feelings associated with drug-induced dopamine release were supposed to be much stronger than those elicited by food consumption or sexual intercourse. In fact, microdialysis studies reveal that psychostimulants including cocaine or amphetamine increase dopamine release by a factor of 6 to 12, while food increases basic dopamine release at most twofold (Di Chiara and Imperato, 1988; Martel and Fantino, 1996; Fuchs, Nagel, and Hauber, 2005).

However, a series of findings questioned this "drug hedonia hypothesis": Wolfram Schultz observed that dopamine is only released if reward is surprising, while there is no dopamine release once the outcome is fully predicted by a conditioned cue (Schultz, Dayan, and Montague, 1997; Schultz 2004; 2007). In his landmark study, the outcome was food (banana chips) acquired by rhesus monkeys. Rhesus monkeys love bananas, and it is unlikely that a fully predicted food reward is no longer pleasurable—if so, the monkeys would most likely cease to work for this reward. However, availability of bananas is no longer associated with dopamine release once it is fully predicted by preceding conditioned stimuli (figure 3.1).

On the basis of this observation, it was suggested that dopamine does not encode the pleasurable effects of reinforcers but rather reflects an error of reward prediction (i.e., dopamine release is increased whenever the acquired outcome is better than expected). Schultz et al. (1997) indeed observed that a cue that fully predicts upcoming reward is accompanied by a phasic increase in dopamine release, which reflects an error of reward prediction—at the moment when the conditioned cue is presented, the animal did not expect the occurrence of a stimulus that predicts reward.

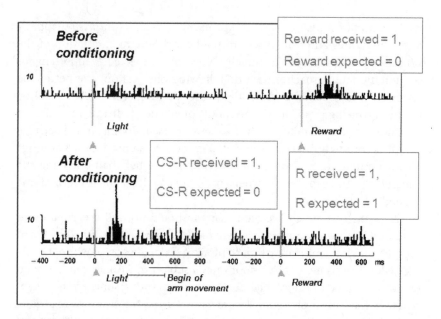

Figure 3.1

Phasic dopamine release encodes errors of reward prediction. Top: Recording of dopamine cell firing before learning that a conditioned stimulus reliably predicts the availability of reward once an arm movement is executed. Phasic dopamine release was elicited by an unpredicted reward before the non-human primate learned that the food reward is announced by a conditioned stimulus (a red light). Because the reward was unpredicted at this time point (Reward expected = 0), the received reward (Reward received = 1) is much larger than expected, and the difference between the expected reward and the received reward (Reward received – Reward expected = 1 – 0 = 1), that is, the prediction error, is positive and elicits a phasic increase in dopamine firing rates. The degree of the phasic increase reflects the size of the prediction error. Bottom: Dopamine cell firing after conditioning. Once the monkey learned that the conditioned stimulus (CS-R, conditioned stimulus predicting reward) reliably predicts the availability of reward when pushing a button, the conditioned cue is valued as positively as the upcoming reward, and its unexpected occurrence elicits a positive prediction error. The fully predicted food reward, in contrast, is not surprising and constitutes no error of reward prediction (Reward received – Reward expected = 1 – 1 = 0) and hence no alteration in basic dopamine firing rates (Schultz, Dayan, and Montague, 2007).

As the stimulus is associated with upcoming reward, its surprising appearance constitutes an error in reward prediction—the manifestation of the stimulus at this moment is "much better than expected." Important for computational approaches was the following observation: the better the expected value of the outcome, the higher the dopamine release elicited by the surprising exposure to the conditioned cue predicting reward or to the unexpected availability of the food reward itself. Robinson and Berridge (1993) accordingly suggested that dopamine release promotes "wanting" (i.e., a positive motivation to acquire reward), rather than encoding the hedonic pleasure experienced when consuming a fully predicted reward (the "liking" of the outcome).

These observations first suggest that we should not call every reinforcer a "reward," at least as long as we assume that a reward is defined by eliciting pleasure. Dopamine may thus reinforce a certain behavior without evoking hedonic pleasure, rather, it motivates a subject to work to gain a certain reward, be it food or drugs of abuse. Dopamine is thus assumed to be associated with the "drive" or, to put it more neutrally, the "motivation" to acquire a certain outcome rather than associated with a pleasant "feeling." These considerations suggest that we stick with traditional behaviorist vocabulary and speak of reinforcers, thus avoiding the implicit assumption that every positive reinforcer is experienced as a subjective reward (i.e., hedonic pleasure). However, these considerations do not rule out that dopamine release itself is associated with positive emotions beyond hedonic pleasure, as for example can be experienced when expecting that one's own activities help to gain a desired outcome. Compare for instance subjective emotions elicited by opioids, a rather quiet and joyful state, which corresponds to hedonic pleasure when consuming a fully expected reward on the one hand, and effects of psychostimulants such as cocaine on the other, which elicit a feeling of power and a desire to act out. As far as the motivation is concerned to act on one's perceived power, dopamine release elicited by psychostimulants appears to play a dominant role. However, other neurotransmitter systems are usually also activated by consuming cocaine or other psychostimulants such as amphetamine; for example, these drugs of abuse also release noradrenaline and serotonin, which appear to mainly increase arousal and positive emotions, respectively (McCormick et al., 1992; McCormick, 1992; Meyer, 2013). These observations suggest that we carefully distinguish between different aspects of positive reinforcement and their potential subjective correlates.

Second, how straightforward is operant conditioning? In fact, we are usually not confronted with overly simple situations in which a certain action automatically leads to a specific reward. Rather, we often have to decide what to do, consider different options, and evaluate the respective choices. A straightforward way to describe reward-related learning mechanisms when confronted with such choices is given by a modified Rescorla-Wagner model, which was originally developed to explain Pavlovian rather than operant conditioning (Rescorla and Wagner, 1972; Sutton and Barto, 1998). Here, the valence of a certain decision depends on the previously learned valence and its update according to the size of the reward prediction error encountered due to the current choice, which is multiplied by the learning rate with which it is scaled:

$$V_{T(0+1)} = V_{T(0)} + \alpha \times (\text{Reward}_{\text{received}} - \text{Reward}_{\text{expected}})$$

where $V_{T(0+1)}$ is the value of a certain choice at the moment (arbitrarily set between 0 = no reward and 1 = full reward), $V_{T(0)}$ is the value of the last choice, and α is the learning rate (between 1 = immediate complete learning and 0 = no learning, determining the rate of change in learning), and $(\text{Reward}_{\text{received}} - \text{Reward}_{\text{expected}})$ is the error of reward prediction.

Imagine that you are confronted with two choices: you can press the button on the left or the right side, just as you can decide to drink or not to drink. At the beginning of your trial, you do not know which choice is better. A full reward is arbitrarily set at one; at the start, you do not know which side to choose, both options are equally likely to offer the reward, hence the value of action before you start to choose is 0.5 for both sides:

$V_0 = 0.5$ $\hspace{8cm}$ $V_0 = 0.5$

The task is a probabilistic reinforcement learning task, meaning that the currently "correct" choice will be rewarded in 80% of all instances it is chosen, while it is punished in the remaining 20% (i.e., you gain no reward). After a while, the winning side will change unexpectedly, and what has been a previously rewarded choice is now punished, so you have to learn again which side to choose (figure 3.2).

Imagine that you are making your first choice, you press the right button and you were lucky, your choice was correct. Hence your new choice values will be as follows:

$V_{T(0+1)} = 0.5$ $\hspace{3cm}$ $V_{T(0+1)} = 0.5 + \alpha \times (\text{Reward}_{\text{received}} - \text{Reward}_{\text{expected}})$

The learning rate α can be individually estimated from the observed data; it often has values around 0.3, which for simplicity we will use in our model. $\text{Reward}_{\text{received}}$ was 1, reward expected was 0.5, and hence your prediction error is $1 - 0.5 = 0.5$. The values of your options accordingly look like this after you made the first choice:

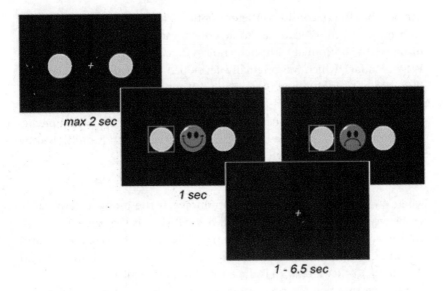

Figure 3.2
Reversal learning task. The proband is confronted with two options, either to press the button on the left side or to press the button on the right side. Choice is indicated by a blue box. At the start, the proband does not know which side to choose. In the example displayed here, choosing the left side is probabilistically rewarded in 80% of all trials, while it is punished in 20%; from time to time, the "winning side" is reversed, and choosing the right side will now be probabilistically rewarded in 80% of all trials.

$$V_{T(0+1)}=0.5 \qquad\qquad V_{T(0+1)}=0.5+0.3\times0.5=0.65$$

When you compare your left- and right-hand choices, the value for choosing the right side now is larger than the value for the left side, so you will most likely choose the right side again (at least if you behave like our simple computing machine, which is meant to model your behavior). Imagine that you are rewarded again; accordingly, the values of your choices now stand as follows:

$$V_{T(0+2)}=0.5 \qquad\qquad V_{T(0+2)}=0.65+0.3\times(1-0.65)=0.755$$

Now imagine you encounter the first probabilistic error, which means that choosing the right side is still generally okay; however, you encountered one of the 20% of all situations where in spite of choosing the correct side, you are punished. Nothing will happen to the value of your left-hand choice; however, by choosing the right side, you encounter a negative prediction error, that is, your expected reward (0.755) is larger than the reward you actually received (0):

$$V_{T(0+3)} = 0.5 \qquad V_{T(0+3)} = 0.755 + [0.3 \times (0 - 0.755)] = 0.755 - 0.2265 = 0.5285$$

Now your right-hand choice is just slightly better than the left-hand choice; however, you will again press the button on the right-hand side (again: if you behave as a computing machine), as it still is better than the choice on the left-hand side (in case you are that sensitive to little differences even though you are not allowed to use a pocket calculator while you act). Imagine now the correct side has switched and it is better to press the left-hand button. Accordingly, you will be punished in 80% of all instances when pressing the right-hand button, and indeed this time you receive a negative feedback (the received reward is 0). The count now stands as follows:

$$V_{T(0+4)} = 0.5 \qquad\qquad V_{T(0+4)} = 0.5285 + 0.3 \times (0 - 0.5285) = 0.36995$$

Accordingly, now the value on left is larger than on the right, and as a rational and exactly computing agent, you switch to the left side. In this case, this is a favorable strategy; however, remember that it could also be that no switching of winning sides occurred and you just accidentally encountered one of the 20% of probabilistically negative outcomes on the right side. Then switching to pressing the left button will be punished rather often, and you have to learn again which side to prefer.

This simple thought experiment is valuable because it can help to explain how a simple computer program can capture and model a behavior strategy depending on reward probabilities and the volatility; that is, the stability or instability of the situation (i.e., rare or frequent switches of favorable choices). If the human volunteer followed exactly the same strategy as calculated and exerted by the computer, then the choices calculated by computing device should be exactly like the choices of the human being. Remember that the human being is not allowed to use a pocket calculator; hence, your volunteer may follow the same strategy as the computing device would suggest, and yet she would deviate slightly from the exact choices generated by the computer. Nevertheless, you can see whether the behavior of your volunteer generally fits with the choices of the computer. If so, then your human behavior is quite successfully "modeled" by the computer, and you know that the same computational steps have likely been taken by the human being and the machine as quantified by a certain probability, the likelihood that the data are given by the parameters of the model. Such reinforcement learning models compute action values as described above and furthermore contain an algorithm determining how one of the choice options is selected on the basis of these action values. One often-used algorithm is the so-called softmax function, which transforms values into action probabilities and incorporates a so-called temperature parameter that determines the degree to which an agent follows the learned values.

Having established that a person follows certain computational steps, you can look for neurobiological correlates of such computational steps; for example, the size of the prediction error is supposed to be reflected in the amount of phasic dopamine release, which you can measure electrophysiologically or via microdialysis in an animal experiment or, as a proxy, via the degree of functional activation of the ventral striatum, a projection area of dopaminergic neurons that is usually activated by prediction errors (O'Doherty et al., 2003). You now have the ability to directly correlate a biological signal with a computational step that is relevant for human behavior, as indicated by the fit between decisions made by your computer model and decisions made by a human being. Instead of relying on subjective responses of your subject ("Did I like the reward? How strong was my positive emotion?"), you now replace subjective reports about personal feelings with a computational model of behavior, which aims at better identifying the exact biological correlates of the relevant computational steps—at least as long as the computing device and the human being perform similar calculations, which you can check by comparing the series of choices generated by your computer model with those performed by your proband.

There are, however, different strategies that you can use to act even in such a simple setting—you may react to punishments more than to rewards; that is, you may weigh them more strongly and change your behavior quicker when a punishment occurred, you may also compute on the potential gains you *could have* won if you *had* chosen the other side (i.e., you "double-update" values on the left and right sides), and so forth. To identify which computational model best captures the behavior of human volunteers performing the same task, you have to compare the different models with respect to how well they simulate behavior patterns of patients and healthy volunteers. As an outcome measure, you can, for example, use the Bayesian information criterion; here, low numbers indicate good model fit. When you assess choice behavior in the rather simple and straightforward reversal learning task, you may be surprised how many strategies can be applied to cope with the challenge. The simple "single-update" (SA) strategy that we have just described, in which you update feedback only for the side that you actually chose, fares worst—it does not fit behavior as displayed by most healthy volunteers who perform the task. Altering the model a little by attributing separate weights to rewards and punishments (SA R/P) is somewhat better. When you construct a strategy that updates feedback on both sides—that is, it computes what you gained by choosing, for example, the left side but also updates values on the side you did not choose (double-update model; DSA), according to what you might have lost—and weighs punishments and rewards separately (DSA R/P), you capture actual behavior of your subjects quite well.

However, probands may also try to cognitively map the task structure and, for example, understand that this is a probabilistic reversal task, in which you should switch as soon as you encounter, for example, two punishments in a row. The probands may therefore start counting and behave according to a cognitive map of the task that is hidden to the outside observer (e.g., a so-called hidden Markov model; HMM). If you combine such a model (the hidden Markov model) with different weights for reward and punishment (HMM R/P), you best explain the observed behavior of your volunteers when coping with some of such tasks, and therefore you can assume that the computations used in your model of behavior were also carried out by at least the majority of your probands. In fact, you can check every single participant to identify whether the best-fitting model actually captures her behavior better than chance (Schlagenhauf et al., 2014) (figure 3.3).

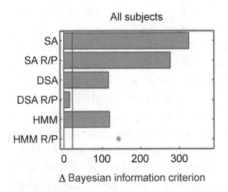

Figure 3.3
Model comparison for the probabilistic reversal learning task. Model comparison is applied to identify the computational model that best fits the behavior of probands coping with the probabilistic reversal learning task. SA, single-update model; DSA, double-update model; HMM, hidden Markov model; R/P, different weights for reward and punishment. Source: Modified according to Schlagenhauf et al. (2014).

Third, what happens with reward learning when you consume drugs of abuse? As indicated above, dopamine release and hence the positive reinforcement associated with drug consumption is stronger than the respective dopamine release elicited by natural reinforcers. Hence, even if drug consumption is fully predicted by conditioned stimuli, the ensuing dopamine release could be stronger than expected, simply due to the pharmacological effects of drugs of abuse on dopamine release.

When considering the Rescorla-Wagner model of operant learning discussed earlier, receiving reward due to a drug of abuse compared to natural reinforcers such as food or sex should be particularly strong and therefore bias decision making toward drug consumption. Consider for example in the earlier equations what would happen if you receive reward due to a drug of abuse, which strongly releases dopamine and accordingly elicits a large error of reward prediction: in the ensuing equations, you would be strongly inclined to favor this reward even though the learning rate may remain constant (α would still be set at 0.3), and your reward expectations will be updated according to the received reward. If for example by pressing the left button, you can receive a natural reinforcer such as food, while when pressing the right button you will receive cocaine, you will be biased toward working for cocaine because of the higher dopamine release and hence the stronger error of reward prediction elicited by the drug of abuse, as the prediction error will be incorporated in your subsequent expectations of the reward value potentially available when consuming either food or cocaine (Redish, 2004; Redish and Johnson, 2007).

Fourth, even though some computational power is involved in calculating the equations listed earlier, reward learning according to a Rescorla-Wagner model is apparently too simple to comprehensively capture human behavior even when confronted with a rather simple probabilistic reversal learning task. It does not require mapping a complex series of decisions; rather, you can act according to the most recently received reinforcement. Reward learning after such rather simple procedures is usually called "model-free" reinforcement learning. This wording may be rather confusing, as subjects indeed follow a Rescorla-Wagner model of decision making, which hence is a *model*, even if a simple one. When choosing such a simple decision-making strategy as in a "single-update" Rescorla-Wagner model, we indeed use a (simple) computational model to analyze behavior; however, the proband herself just reacts on the most recent feedback and does not cognitively model the task structure. This is why the term "model-free" is applied to her behavior, which indicates that the person does not model all options in the so-called decision tree but instead allows her behavior to be driven by just

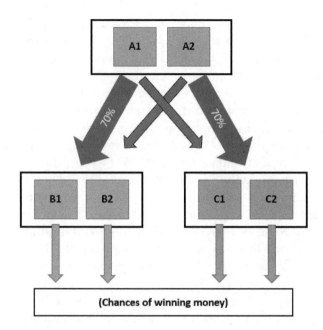

Figure 3.4
Two-step task. In this task, you have to make two consecutive choices: on the first level, you choose to press the left- or right-hand button, and you will probabilistically be led to a second level, in which you have to repeat choices. For example, when pressing the left button (A1), you will frequently (i.e., in 70% of all instances) be led to the choice between B1 and B2, while in rare cases (i.e., in 30%), you will unexpectedly be confronted with the choice between C1 and C2. On the second level, choices are rewarded with slowly and randomly changing fluctuations to increase volatility at this choice level.

one "branch" of that "tree"—the most recent feedback. Also, by reacting to the most recent feedback, model-free behavior is "retrospective," while model-based behavior takes different options "prospectively" into account.

More complex modeling of a complete decision tree can be required if you have to go through a series of at least two choices (figure 3.4): on the first level, you can decide whether you press a left or right button and with some probability you are led to a second level of choices, on which you have to repeat choosing the left- or the right-hand button (Dolan and Dayan, 2013; Akam, Costa, and Dayan, 2015).

When confronted with this task, you can follow a "model-free" way of decision making by just orienting your responses according to whether you most recently received a reward or not. In case of being rewarded, you just repeat the choices, independent of whether you encountered a frequent and hence more likely transition at the first level or not. If you act like this (i.e., if your behavior is driven by the most recent reward without taking the complex task structure into account), you will repeat any series of choices rewarded, while any series of choices that was punished will not be repeated. However, if you take task structure into account and map it by creating a model of the task, you can act "model-based." When doing so, you will consider that on the first level, in 30% of all choices you will encounter a rare and hence rather unexpected transition to the next level of choices. Therefore, if you understand that you were just rarely led to a different set of choices, you can include this knowledge into your decision making and, for example, repeat the choice on the first level because it will most likely lead you next time to the set of second-level choices that you already known quite well, which will hence allow you to make a successful choice and gain a reward on that second level. Most important, you will repeat the choice on the first level even if it resulted in a rare and hence unexpected transition to the second level of choices, where you made a mistake and were punished. According to operant learning theory, encountering punishment usually tends to inhibit repeating a choice—to act "model-based," you have to overcome this bias and orient your behavior at the cognitive insight that you just encountered a rare situation in which a usually rewarding choice was accidentally not successful (figure 3.5).

We will discuss the mathematical equations associated with such more complex, "model-based" ways of decision making in a while. At this point, it may suffice to understand that reward-related decision making can be rather straightforward or very complex: as in real life, your choices may not immediately offer you reward, but instead they result in further options, and whatever choices you make on early levels of decision making, you will most likely not be sure that they offer you exactly the desired next set of choices.

Model-free decision making, which just relies on the most recently offered reinforcement, has also been called habitual, while model-based decision making, which requires mapping of all options of a certain "decision tree," has also been called goal-directed, suggesting that model-based mechanisms describe the computations underlying the flexible and complex pursuit of goals in volatile environments. In fact, a behavior is usually called goal-directed when the individual actions that together constitute the observable behavior are modified by the desired goal. For example, you may act according to a certain task and solve a series of riddles if you desire a certain food reward (e.g., chocolate). To test whether this goal is still

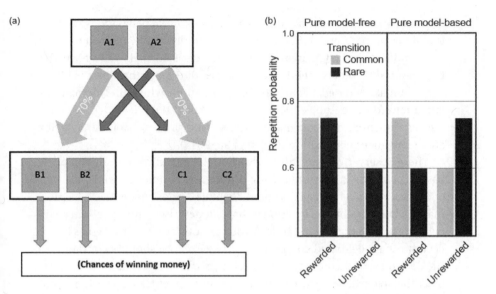

Figure 3.5

Model-free versus model-based decision making in the two-step task. (a) Task structure: On level 1 a side is chosen, and the proband will be probabilistically led to level 2, where the proband encounters new choices, which will then finally be rewarded. (b) Left panel: Model-free decision-making pattern. Choices on level 1 are repeated whenever the second choice on level 2 was rewarded, independent of whether the proband encountered a rare or frequent transition on level 1. Right panel: Model-based decision making. A choice on level 1 is not repeated, even though it led to a second level of choices on level 2 that was finally rewarded, because the proband understands that repeating the choice on level 1 will more likely (70%:30%) lead to a different set of options on level 2. Also, a choice will be repeated even though the proband encountered a rare transition on level 1 (30% chance) and was unexpectedly confronted with a second level of choices and lost.

driving your behavior, you are then offered as much chocolate as you want to eat. Once satiety is reached, the previously desired reward (chocolate) has lost its positive value. If you now stop to work for chocolate and, for example, instead work for orange juice, your behavior will be called "goal-directed," as it was influenced by the devaluation of the previous pursuit goal, chocolate. If you continue to work for chocolate, although satiety has been reached, and this goal is therefore no longer valuable, your behavior will be called "habitual."

Accordingly, to decide whether a behavior is "model based" or "model free," rather different experimental procedures are applied than when

distinguishing between goal-directed versus habitual behavior. Nevertheless, it has been observed that in case you are individually inclined to show
more model-based compared to model-free decision making (e.g., in the
two-step task), you also tend to display goal-directed rather than habitual
decision making (Friedel et al., 2014). Furthermore, when directly manipu-
lating reward value, subjects who strongly displayed model-based behavior
during sequential decision making indeed showed less habitual behavior
after subsequent devaluation (Gillan et al., 2015).

These observations suggest that although "goal-directed" and "model-
based" behavior is measured using different experimental tools, it often
coincides in a given subject, as does "habitual" and "model-free" decision
making. One possible explanation is that "model-based" and "goal-directed"
behaviors are the same, and indeed you cannot flexibly pursue goals if you
do not use model-based computations; however, we should be aware that
the definition of theoretical constructs such as goal-directedness or model-
based behavior depends on how they are experimentally operationalized.
Furthermore, when looking at the behavior of individual subjects, we talk
about different degrees of habitual versus goal-directed behavior, not about
categorical differences between individual subjects.

When we consider Pavlovian influences on automatic reactions, we will
also tend to think that automatic responses such as saliva production when
confronted with a Pavlovian conditioned cue is rather habitual and independent of creating a cognitive model or map of the world. In other words,
a dog will produce saliva once a bell rings independent of whether there
is some cognitive representation of the bell sound and the food—the association is supposed to be automatic or even "hardwired" in the biological
makeup of the animal. So again, we can ask whether Pavlovian-to-instrumental transfer as described in chapter 2 is more powerful in individuals
who display model-free rather than model-based behavior in the two-step
task. Indeed, this appears to be the case (Sebold et al., 2016): subjects who
show model-free behavior in complex decision-making tasks tend to be
strongly influenced in their instrumental choices when confronted with
unrelated Pavlovian background cues, while this is not the case in subjects
who mainly display model-based decision making. However, finding associations between these different ways of responding to reward and punishment with respect to individual differences again does not prove that the
underlying mechanisms are identical. Rather, it suggests that as humans, we
can be biased toward responding rather automatically to Pavlovian stimuli
and to immediate rewards or we can tend to calculate more carefully all
options and to act on such computations.

It is easy to understand that complex calculations have internal "computing" costs: keeping track of two levels of choices in the two-step task demands working memory capacity, and individual differences in working memory performance are associated with model-based versus model-free decision making (Schad et al., 2014). Also, when you are busy calculating options, your attention is focused and you may not be able to spontaneously react to other challenges; therefore, a dual-task challenge reduces the degree of model-based behavior (Otto et al., 2013a). Acting model-based, you will most likely only be able to focus on one specific problem rather than be able to cope with a series of simultaneous challenges. Finally, whether you are able to compute complex calculations and to act according to them may depend on your level of wakefulness and attention. Indeed, many subjects use short breaks in order to refocus their attention when they get tired, and break activity has been shown to influence the way we cope with complex decision-making paradigms (Liu et al., 2016). This latter observation suggests that rather than being "hardwired," model-based and model-free ways of decision making depend on a series of contextual factors, not the least being internal influences such as wakefulness and attention. We emphasize these points because one may assume that goal-directed behavior is always preferable and "more rational" than habitual decision making. However, as human beings, we need automatic behavior: presumably the majority of our actions are habitual, and goal-directed actions are only required when habits fail to cope with volatile environments or when new challenges appear. Therefore, we suggest that both approaches have their respective advantages and limitations, and they may lead to different mental disorders.

Before we try to apply the above-listed learning mechanisms to clinical cases, in the next chapter we will describe the computational structure of model-based versus model-free choice in more detail, because some paradigms aiming at these different ways of decision making have been applied to better understand clinical disorders such as obsessions, compulsions, and addictions (Voon et al., 2010; Sebold et al., 2014).

4 Executive Control and Model-Based Decision Making

At this point, we should look at the computational structure of the two-step task to understand better which steps are involved in model-based versus model-free decision making. At the second level of decision making, when choices are directly rewarded or not, the computational structure of the two-step task is in accordance with the structure of model-free learning as described with respect to the reversal learning task in the past chapter. This means that reward values are directly updated depending on the last choice according to a prediction error that is weighted via an individual learning rate. Equations implicated in this task can be described as follows when extending to a scenario with two subsequent choices as in sequential decision making.

Model-free action values (Q_{MF}) are learned via continuous updating using prediction errors that signal the difference between expected and received outcome value (Sutton and Barto, 1998):

$$Q_{MF}(s_{i,t+1}, a_{i,t+1}) = Q_{MF}(s_{i,t}, a_{i,t}) + \alpha_i \delta_{i,t}$$

Here, i is an index for the states s observed and the actions a performed in the sequential decision-making task (first or second step/stage), and t indexes time in terms of trials. Thus, $Q_{MF}(s_{i,t}, a_{i,t})$ represents the model-free "state-action value" of being in state s on level i in trial t and choosing action a; α is the learning rate on level i, which weighs the prediction error δ on level i in trial t.

Prediction errors δ happen at the onset of the second stage and at reward delivery after a second-stage choice:

$$\delta_{i,t} = r_{i,t} + Q_{MF}(s_{i+1,t}, a_{i+1,t}) - Q_{MF}(s_{i,t}, a_{i,t})$$

Here, $r_{i,t}$ describes the reward obtained at level i in trial t. At the first stage $r_{i,t} = 0$, because rewards are delivered at the second stage only. After the first choice on stage 1, the prediction errors at the onset of the second stage reflect the difference between two state-action values, the value at the second stage, $Q_{\mathrm{MF}}(s_{2,t}, a_{2,t})$, minus the value at the first stage, $Q_{\mathrm{MF}}(s_{1,t}, a_{1,t})$. Note that there is no action value for $i=3$, that is, the action value for the third level $Q_{\mathrm{MF}}(s_3, a_{3,t}) = 0$, because there are only two stages. The prediction error at the second stage is thus equivalent to prediction errors as described before with respect to reversal learning, $\delta_t = r_t - Q(a_t) = r_{2,t} - Q_{\mathrm{MF}}(s_{2,t}, a_{2,t})$.

The computationally more complex model-based decision making, in contrast, relies on representing the probabilistic structure of the first choice (i.e., the subject needs to understand whether the currently experienced transition is a frequent or rare one). Of course, it makes sense to repeat first-level choices when they are frequently rewarded, but, as described in the previous chapter, it also pays off to repeat choices that are rarely punished, because repeating a choice on the first level with a rare transition and ultimate punishment will most likely this time lead to a more frequent transition with choices that can be successfully mastered. Thus, to allow learning of model-based state-action values, we need to focus on how to modify Q at the first stage ($i=1$ or S_A in the following) to promote behavior reflecting the transition structure. This is represented in the above listed equations as follows:

$$Q_{\mathrm{MB}}(s_A, a_j) = P(S_B|S_A, a_j)\, \max Q_{\mathrm{MF}}(s_B, a) + P(S_C|S_A, a_j)\, \max Q_{\mathrm{MF}}(s_C, a)$$

The three possible states of this particular two-step task are denoted with state S_A being in the first level and with S_B or S_C being in one of the second-level states. $Q_{\mathrm{MB}}(s_A, a_j)$ represents the value of the model-based action a_j at the first stage S_A. $P(S_B|S_A, a_j)$ is the probability to be in state S_B (one of the two possible second-level states) after action a_j has been taken at the first-level S_A. Thus, $P(S_B|S_A, a_j)$ and $P(S_C|S_A, a_j)$ take numerical values of 0.7 or 0.3, respectively, depending on the type of transition (common = 70% probability or rare = 30% probability). The term $\max Q_{\mathrm{MF}}(s_B, a)$ represents the currently maximal value out of two choice action (a) values for the stimulus pair S_B, and $\max Q_{\mathrm{MF}}(s_C, a)$ for the stimulus pair S_C. Please note that this multiplication of transition probabilities, $P(S_B|S_A, a_j)$ and $P(S_C|S_A, a_j)$, with second-stage model-free values, $\max Q_{\mathrm{MF}}(s_B, a)$ and $\max Q_{\mathrm{MF}}(s_C, a)$, simplifies model-based learning but enables model-based control in this certain version of sequential decision making, in which the transition structure of the task is excessively trained before the main experience (compare Daw et al., 2011). For incremental model-based learning, please see Glascher et al. (2010) and Lee et al. (2014).

When we look at the neurobiological correlates of model-based and model-free decision making, it has been shown that both involve functional

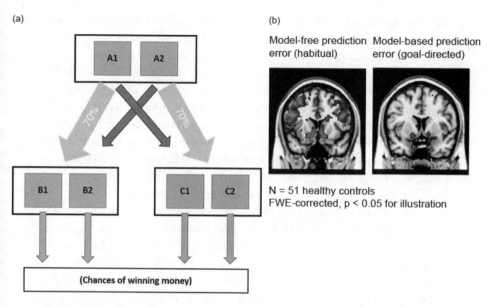

Figure 4.1

Activation of the ventral striatum during both model-based and model-free decision making. (a) Structure of the sequential decision-making task (two-step task). (b) Left panel: Functional activation elicited by model-free prediction errors associated with model-free decision making activates the ventral striatum and frontal cortex. Right panel: Additional activation of the ventral striatum is observed during model-based decision making (Deserno et al., 2015). FWE, family-wise error.

activation of the prefrontal cortex but also, and maybe less expected in traditional accounts, the ventral striatum (figure 4.1) (Daw et al., 2011; Dolan and Dayan 2013; Deserno et al., 2015a).

This may be surprising, because model-based decision making is supposed to be associated with complex cognitive planning and mapping of the task structure; accordingly, it requires cognitive resources including working memory capacity and cognitive processing speed. Indeed, Schad and coworkers (2014) observed that cognitive processing speed measured in an unrelated task (the digit symbol substitution task) is linearly and positively correlated with model-based decision making in the two-step decision task described before. Notably, model-based decision making was most pronounced in individuals with fast processing speed and high working memory performance (Schad et al., 2014). A link between model-based decision making and individual differences in working memory capacity as well as further cognitive control functions including the utilization of goal-directed contextual information was confirmed in additional independent studies (Otto et al., 2013b; Otto et al., 2015). Executive functions such as working memory have repeatedly been associated with activation of and connectivity between prefrontal and parietal cortical brain areas (Deserno et al., 2012; Ester, Srague, and Serences, 2015). Therefore, one might expect to find cortical, particularly prefrontal cortical, brain activation to be involved in model-based learning and decision making, which has indeed been demonstrated (e.g., Glascher et al., 2010). However, individual differences in functional activation of the ventral striatum elicited by reward prediction errors has been associated with fluid intelligence (Schlagenhauf et al., 2013), indicating that a key aspect of higher cognitive functioning (i.e., processing speed or so-called fluid IQ) correlates with information processing in a subcortical brain area, the ventral striatum. Similar findings have been revealed during sequential decision making (Daw et al., 2011; Deserno et al., 2015a) and also reversal learning (Hampton et al., 2006; Schlagenhauf et al., 2014), pointing toward a role of the ventral striatum in higher-order cognitive control processes. How can such findings be explained?

In chapters 6 and 7, we will address hierarchical and non-hierarchical models of brain functioning in more detail. At this point, we already like to emphasize that the brain functions as a highly interactive network structure, with top-down and bottom-up information processing as well as parallel processing of information in parallel neurocircuits. An important subgroup of such neurocircuits links prefrontal cortical brain areas with different parts of the ventral and dorsal striatum and the thalamus and thus contributes to motivational and motor processes (Alexander, DeLong, and Strick, 1986; Fuster, 2001). It would therefore be wrong to assume that higher cognitive functioning relies exclusively on cortical information processing, while automatic information processing depends solely on subcortical brain functions, as may be assumed when model-based versus model-free decision making is reified into a "dual systems theory." This latter theory suggests that there is a balance between two separate "systems" of decision making, one dedicated to complex mapping and representation of all choice options (the "model-based system") and the other relying on habits and thus only using a certain part of the decision tree (the "model-free system"). However, empirical data suggest that subcortical brain areas such as the striatum are involved in both higher-order model-based as well as less complex model-free decision making, most likely due to the striatum not acting on its own but being embedded in neurocircuits that link cortical and subcortical brain areas in order to facilitate decision making at different levels of complexity.

Another caveat points to the fact that functional brain activation, as observed when computing prediction errors and correlating them with the blood oxygen level–dependent (BOLD) signal measured by functional magnetic resonance imaging, does not directly assess dopaminergic functioning. In fact, functional brain signals reflect mainly the energy demand of nerve cells associated with information input (rather than output) (Logothetis et al., 2001). Given the observation in animal experiments that the amount of phasic dopamine release corresponds to the size of a prediction error (Schultz et al., 1997; Schultz, 2007), it is plausible to assume that dopaminergic neurotransmission plays a role in the striatal computation of such prediction errors and hence in altering neural network activation associated with information input to a degree that can be measured by functional magnetic resonance imaging. However, to confirm such hypotheses in humans, functional magnetic resonance imaging has to be combined with techniques that directly measure individual neurotransmitter concentrations or release, such as positron emission tomography, and pharmacological manipulations can also be useful. Indeed, Schott and coworkers (2008) observed that phasic alterations in dopamine release measured indirectly via displacement of a radioligand from dopamine D2 receptors in the ventral striatum, a key projection area of midbrain dopaminergic neurons, is positively correlated with the degree of functional activation elicited by reward anticipation (i.e., the unexpected appearance of a cue indicating available reward and thus inducing an error of reward prediction at the time it appears) as measured with magnetic resonance imaging in the substantia nigra and ventral tegmental area in the midbrain, where dopaminergic neurons originate. However, the role of dopamine in regulating prediction errors in prefrontal-striatal-thalamic networks appears to be more complex: Wunderink and coworkers (2013) observed that elevation of overall presynaptic dopamine levels (via application of dopamine precursor levodopa; L-dopa) leads to enhanced model-based decision making. Deserno et al. (2015a) extended this initial finding by the observation that in the ventral striatum, presynaptic uptake of the same dopamine precursor radiolabeled as F-dopa, an index of long-term dopamine synthesis capacity, is positively correlated with the degree of model-based choices as well as functional activation of the lateral prefrontal cortex associated with model-based control. This finding points to a direct interaction between a marker of presynaptic dopaminergic neurotransmission in the ventral striatum (F-dopa uptake) and information processing in a frontal cortical brain area associated with cognitive control. Also, model-based (but not model-free) control was found to be impaired in Parkinson's patients, and this effect was selectively remediated with dopaminergic drug therapy (Sharp et al., 2016).

These findings are in accordance with hypotheses that assume a "gate-keeper" role for the striatum with respect to cortical information processing and the selection of behavioral strategies (Cohen and Servan-Schreiber, 1992; Goto and Grace, 2008). Behaviorally, increased model-based control is usually associated with less reliance on habitual, model-free decision making and vice versa (Doll et al., 2016). In accordance with that assumption, Deserno et al. (2015a) observed that dopamine synthesis capacity in the ventral striatum was negatively correlated with coding of model-free prediction errors in the striatum, thus replicating a finding of Schlagenhauf and coworkers (2013).

Altogether, these observations emphasize the importance of dopaminergic neurotransmission in the striatum for model-based as well as model-free decision making. With respect to model-based decision making, the evidence also suggests an important role of dopaminergic neurotransmission in other brain regions including the prefrontal cortex, as suggested, for example, by a study assessing genetic variation in the dopamine-metabolizing enzyme catechol-O-methyltransferase (COMT), which mainly affects extracellular dopamine concentrations in the frontal cortex and further brain areas mainly devoid of dopamine transporters (Doll et al., 2016). Altogether, the available evidence promotes an integrative view of how behavioral control emerges through spiraling fronto-striatal loops. This view is also supported by the observed links between working memory performance and markers of prefrontal (Williams and Goldman-Rakic, 1995; Abi-Dargham et al., 2002) as well as striatal (Cools et al., 2008) dopaminergic neurotransmission.

Working memory capacity and other parameters of cognitive performance (including fluid IQ) do not appear to be the only factors that influence the balance between model-based and model-free decision making. Also, acute stress exposure interferes with model-based decision making: acute stress reduces model-based choices when working memory capacity is low (Otto et al., 2013b). Another study showed that individuals with high—but not with low—chronic stress levels decreased model-based behavior after exposure to acute social stress (Radenbach et al., 2015). Notably, high levels of chronic life stress were also associated with a stronger correlation between fluid intelligence on the one hand and functional activation associated with model-free encoding of reward prediction errors in the ventral striatum on the other (Friedel et al., 2015). Finally, and again depending on working memory capacity, what leisure activity you choose during a break can also influence your ability to rely on model-based decision making and instead promote model-free, more habitual ways of coping with challenges (Liu et al., 2016).

These observations confirm that the interaction of different brain areas in complex neurocircuits and neural networks underlies task performance during more goal-directed or more habitual ways of decision making. Indeed, considering the role of dopamine functioning during the encoding of prediction errors, it is plausible that the activation of a limited number of dopaminergic neurons in the midbrain can *encode* the size of a prediction error; however, *comparing* incoming information about current rewards with expectations most likely relies on a much wider neural network and demands processing of sensory cues, assessing the values associated with such environmental stimuli, comparing such values with reward expectations according to previous experiences, constructing more or less complex decision trees, and selecting between different behavioral strategies, including more complex goal-directed or more simple habitual ways to respond to a certain situation. In line with this idea, an animal study observed that error signaling by midbrain dopamine neurons reflects the model-based value of cues, thus suggesting access to a wider range of information than provided by model-free learning (Sadacca, Jones, and Schoenbaum, 2016).

Parallel information processing in multiple fronto-subcortical loops (Alexander et al., 1986) may help to explain why even simple choice tasks can be performed using a variety of strategies (Schlagenhauf et al., 2013). Most likely, different strategies are computed in parallel in the brain, with certain approaches winning out against others, depending on individual and situational differences in working memory capacity, uncertainty and ambiguity, but also vigilance and previous experiences (Daw, Niv, and Dayan, 2005; Rangel, Camerer, and Montague, 2008; Adams, Huys, and Roiser, 2016; Liu et al., 2016).

With respect to neurotransmitters involved in these complex information processes, a variety of transmitter systems beyond dopamine has to be addressed. Fast information processing in the brain mainly relies on glutamate as an excitatory neurotransmitter and gamma-aminobutyric acid (GABA) as the main inhibitory transmitter. Dopamine, like the other monoaminergic neurotransmitters noradrenaline and serotonin, arises from the brain stem and modulates fast information processing, mainly by acting on rather slow second-messenger mechanisms that regulate (among others) glutamatergic and GABAergic neurotransmission (Cummings, 1993). Glutamatergic input in the midbrain in turn appears to stimulate dopamine release, while GABAergic neurotransmission, mediated by GABA-B receptors in the midbrain, directly inhibits dopamine discharge in projection areas (Taber, Das, and Fibiger, 1995; Lokwan et al., 2000). To use a reductionist image, the brain may be compared with an old black-and-white television set, in which dopamine plays a role equal to the button that enhanced contrast on the screen. Of course, the brain is a living organ that is much more complex than a television set or any computer humans can construct so far; nevertheless, the television metaphor may help toward understanding that rapid information processing (i.e., *what* is displayed on the television screen) does not directly depend on dopaminergic neurotransmission; rather, higher dopamine release increases the signal-to-noise ratio and hence the level of contrast (i.e., *how* something is displayed on the screen). Remember that in old times, when there was a lot of white noise ("snow") on the television screen and the picture was hard to see, increasing the contrast may have helped to better represent the incoming signal, but at the expense of vanishing details and decreased subtle distinctions, a problem we will return to when discussing the development of psychotic disorders and specifically delusional mood.

While phasic and tonic dopamine release appear to play a role in the encoding of errors of reward prediction, noradrenaline is supposed to increase arousal, and serotonin function has been implicated in the processing of threatening stimuli (Heinz et al., 2011). A substantial part of dopaminergic projections ends in the ventral and dorsal striatum, with the ventral striatum being more involved in motivational processes and attribution of salience toward environmental cues, the central striatum integrating information from a wide range of brain areas, and the dorsal striatum mainly implicated in motor functions. Accordingly, in Parkinson's disease, which is mainly characterized by motor deficits such as akinesia, dopamine dysfunction is largely limited to the dorsal striatum (Rinne et al., 1995).

With continued practice, individuals progress from goal-directed behavior to habits, and information processing in the basal ganglia appears to be shifted from the ventral toward the dorsal striatum: the nucleus accumbens shell of the ventral striatum is supposed to influence activation in the nucleus accumbens core, the core will then influence information processing in the central striatum, and the central striatum finally affects information processing in the dorsolateral striatum (Haber, Fudge, and McFarland, 2000). Dopaminergic projections to the prefrontal cortex, in contrast, appear to be highly important in sustaining information during working memory processes (Williams and Goldman-Rakic, 1995; Goldman-Rakic, 1996).

Because of the role of dopamine in the ventral striatum, this brain area has been identified as a key region of the so-called brain reward system. However, we again would like to emphasize that no brain area operates in isolation; rather, the just described loops between the midbrain and ventral, central, and dorsal parts of the striatum as well as the neurocircuits integrating striatal function into cortical-striatal-thalamic networks all contribute to different aspects of behavior. Furthermore, not all dopamine neurotransmission appears to be related to the encoding of reward-related information: in the amygdala, dopamine synthesis capacity was implicated in processing of aversive but not positive stimuli (Kienast et al., 2008). Again, it would be wrong to limit the complex role of the amygdala to its function in the processing of threatening or aversive cues: in all these brain areas, neurotransmitter functioning has to be understood by elucidating the computational role it plays in the processing of specific information. Our knowledge of such computational roles is currently most advanced with respect to dopamine; nevertheless, dopaminergic neurotransmission never acts alone, and we will have to describe the role of different neurotransmitter systems as well as pharmacological interventions aiming at such systems in more detail when we address clinical phenomena in the next chapters.

Clinically, there is evidence that model-based decision making is impaired in a series of mental disorders that are all characterized by habitually repeating what once has been successful in spite of changing environmental circumstances. This may particularly be problematic when using drugs: when we first consume alcohol, we may enjoy its relaxing or stimulating qualities; first consumption of psychostimulants may also induce a feeling of strength and self-confidence, while consumption of opioids increases joy and decreases pain. However, what mainly were pleasurable effects experienced when first consuming a drug of abuse can end up being disastrous for the individual; for example, due to the neurotoxic effects of alcohol intake or the social and legal problems associated with the acquisition of illegal drugs. Habits of drug taking may thus become disastrous, yet are maintained in spite of their aversive consequences.

One way to explain such behavior patterns is to refer to impairments of model-based choice, and indeed, reductions in model-based decision making have been observed in some but not all studies that assessed patients suffering from substance abuse (Voon et al., 2010; Sebold et al., 2014). Moreover, Voon et al. (2010) suggested that model-based decision making is also impaired in obsessive-compulsive disorders; indeed, drug craving has often been compared to obsessive thoughts, which are experienced by a subject who suffers from such urges without being able to control them, and drug intake in spite of the cognitive decision to remain abstinent has been compared to "compulsive" behavior. When we discuss drug dependence, we will critically review such hypotheses in more detail. At the moment, it may suffice to point to key differences between obsessive-compulsive disorder (OCD) and addiction; for example, the fact that compulsions in OCD usually arise to cope with unpleasant obsessions, while in drug addiction, drug consumption—at least at the onset of drug use—is usually rewarding. Moreover, compulsions in OCD are not mere habits: the subject is not only inclined to continue drug intake automatically but also feels unable to suppress repetition of certain actions for more than a moment, while in drug dependence, excessive drug intake may occur rather automatically (e.g., smoking one cigarette after the other), and craving tends to occur only at certain moments; for example, when the cigarette box is empty and no vending machine is available (Tiffany and Carter, 1998; Schoofs and Heinz, 2013). Such differences are strikingly observable in case a patient suffers from both OCD and drug addiction. In spite of such differences between habits and compulsions, impairments of goal-directed decision making may play a role in a series of mental disorders. With respect to neurotransmitter correlates, beyond dopamine the serotonergic neurotransmitter system has been implicated: reducing serotonergic neurotransmission by consuming a certain diet that depletes tryptophan, the serotonin precursor, impaired model-based decision making during the reward condition but increased it during punishment (Worbe et al., 2015). This observation is in accordance with serotonin playing a role in moderating the effect of aversive or threatening cues, with reduced levels of serotonin turnover and correlated alterations in serotonin transporter availability being associated with stronger amygdala activation and higher levels of aversive feelings such as anger or anxiety (Heinz et al., 2011). Such findings again emphasize that neurotransmitter functions and dysfunctions cut across established nosological boundaries and appear to play a role in several mental disorders, in accordance with the dimensional approach to their neurobiological correlates.

5 Serotonergic Neurotransmission and Its Role in Negative Affect

The serotonin system has often been understood as a counterpart to the dopaminergic system (Daw, Kakade, and Dayan, 2002). In fact, serotonin increases local effects of dopamine on autoreceptors in the ventral tegmental area (VTA) and thus inhibits the firing rate of dopaminergic neurons, which mainly project to the ventral striatum (Brodie and Bunney, 1996). In accordance with this observation, acute blockade of serotonin reuptake, hypothetically increasing acute serotonin concentrations, reduced striatal dopamine concentrations measured in vivo with microdialysis as well as indirectly using positron emission tomography, while blockade of a subtype of serotonin receptors with so-called 5-HT2 receptor antagonists increased extracellular dopamine concentrations in the striatum (Dewey et al., 1995).

Both the dopamine and serotonin system originate in the brain stem and mainly affect postsynaptic second messengers, thus modulating information processing rather than altering it by activating fast-acting ion channels; however, there is an abundance of serotonin receptors, and at least one of them (the 5-HT3 receptor) activates an ion channel, unlike the other receptor subtypes (figure 5.1) (Baumgarten and Grozdanovic, 1985; Tecott and Julius, 1993).

As discussed before, dopaminergic neurotransmission is mainly implicated in reinforcement learning and reward processing, and one of the major types of pharmaceutical drug used in psychiatry, so-called neuroleptics or antipsychotics, directly block dopamine D2 receptors, thus interfering with dopamine's role in attributing a salience to reward-indicating stimuli and potentially impairing motivation (Robinson and Berridge, 1993; Heinz et al., 1998b). Serotonergic neurotransmission, in contrast, is directly affected by many antidepressant pharmaceutical drugs: most of the currently used pharmaceutical drugs block serotonin transporters and hence reuptake of serotonin from the extracellular space (Anderson et al., 2005). Accordingly, serotonergic neurotransmission appears to play a role in the treatment

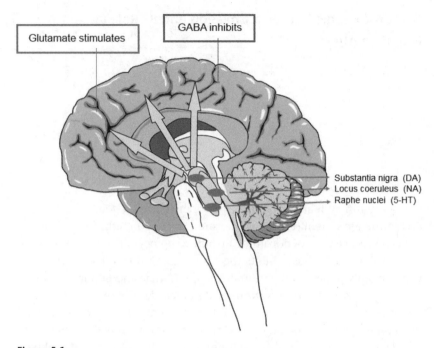

Figure 5.1
Monoaminergic systems modulate neural excitation and inhibition. The monoaminergic systems (dopamine [DA], serotonin [5-HT], and noradrenaline [NA]) originate in the brain stem and modulate excitatory glutamatergic and inhibitory GABAergic neurotransmission in the central nervous system.

and hypothetically also the development of negative mood states, while dopaminergic neurotransmission is rather implicated in positive affective states associated with the anticipation of natural or drug reinforcers (Heinz et al., 2001; Heinz, 2002a).

A reductionist yet very practical approach to the classification of emotions maps them according to valence and arousal, resulting in two major dimensions, one representing positive affect and reaching from excitingly positive to boringly negative, and the other termed negative affect and spanning from highly arousing aversive to boringly positive emotions (figure 5.2) (Watson, Clark, and Tellegen, 1988; Russel, Weiss, and Mendelsohn, 1989).

In accordance with this distinction, antidepressants that block serotonin transporters are mainly used to treat disorders characterized by negative affect such as anxiety or clinical depression. However, it has to be critically

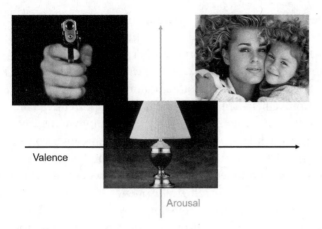

Figure 5.2
Positive and negative affect: mapping affective stimuli according to valence and arousal. Emotions can be mapped according to their degree of arousal and their valence, resulting in an axis of positive affect and another one of negative affect. Peter Lang (1995) identified pictures that elicit emotions in accordance with this hypothesis. As illustrated using such pictures from the International Affective Picture System (IAPS; Lang, 1995), the axis of positive emotions ranges from pleasantly arousing (as elicited by the picture in the right upper corner) to neutral (center) to boringly negative (it is hard to find pictures that are very boring yet very aversive); the axis of negative emotion ranges from highly arousing and aversive (left upper corner) to neutral (center) and to boringly positive (again, it is hard to find pictures that elicit very positive feelings but are quite low in arousal).

remarked that clinical depression may be caused both by an increase in negative mood states and a decrease in positive mood states; for example, neuroleptics are known to induce depression in several subjects treated with antipsychotic medication (Helmchen and Hippius, 1967), and this effect is most likely due to blockade of dopaminergic neurotransmission and its role in encoding the anticipation of reward; that is, it is based on interfering with the encoding of positive expectations rather than by directly promoting aversive mood states (Heinz, 2002a).

How is serotonergic neurotransmission supposed to interact with negative mood states such as anxiety and depression? That a pharmacological drug such as a selective serotonin reuptake inhibitor improves negative mood does not prove a causal role of serotonergic neurotransmission in the origin of aversive mood states. This is clear when we consider the role of anticholinergic medication in Parkinson's disease: a loss of dopaminergic neurons in the substantia nigra of Parkinson patients is associated with a lack of dopamine release in the dorsal striatum, and this dopaminergic dysfunction can be improved by applying anticholinergic medication, which is then impairing a second neurotransmitter system that balances dopamine effects in the striatum. In this case, two wrongs make one right, and a Parkinson's patient treated with an anticholinergic drug will be able to show a clinical improvement of her movement disorder (Streifler et al., 1990). Likewise, blockade of serotonin transporters may improve negative mood states simply because it interacts with other, more causally involved systems or neurotrophic factors (Santarelli et al., 2003; Gulbins et al., 2013). Therefore, to prove a role of serotonergic neurotransmission in the development of negative mood states, neurobiological evidence is needed, which directly shows how serotonin neurotransmission affects neural networks and computations relevant for the processing of affective stimuli and the associated mood states.

One case in point is given by studies looking at the direct effect of drugs of abuse that substantially increase serotonergic neurotransmission. For example, a drug of abuse called MDMA (3,4-methylenedioxymethamphetamine), also known as ecstasy, directly increases serotonin release and thus appears to cause the positive feelings associated with use of this drug (Meyer, 2013). Likewise, cocaine releases not only dopamine but also noradrenaline and serotonin, and cocaine application reinforces further drug consumption even in animals with a dopamine transporter knockout, most likely due to the remaining effects on serotonin transporter function (Sora et al., 2001).

A caveat is that such general effects of drugs of abuse on serotonin transporters may wrongly be understood as simple increases or decreases of extracellular serotonin concentrations. In fact, the effects of serotonin transporter blockade are more complex: elevations in extracellular transmitter levels due to blockade of serotonin reuptake immediately act on autoreceptors, and the time course of alterations of extracellular serotonin concentrations varies considerably in different brain areas (Bel and Artigas, 1993). Indeed, application of antidepressants increases extracellular serotonin concentrations only modestly, with some studies even showing no significant alteration (Smith et al., 2000; Anderson et al., 2005). Moreover, serotonin alterations elicited by blockade of serotonin reuptake act on an abundance of postsynaptic receptors that have quite different functions (figure 5.3) (Tecott and Julius, 1993; Heinz et al., 2011).

Therefore, it is important to look at studies that deplete rather than increase serotonergic neurotransmission and examine whether chronic

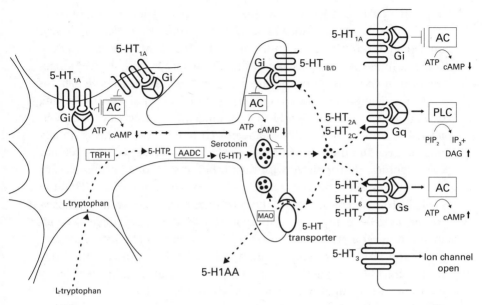

Figure 5.3
Serotonergic neuron and synapse with presynaptic and postsynaptic receptors.
Source: Modified according to Heinz et al. (2012b).

reductions in this neurotransmitter system promote anxious or depressed behavior. Indeed, in a series of studies in non-human primates, the group of Higley and colleagues (1996a, 1996b) observed that social isolation stress experienced directly after birth by rhesus monkeys impairs serotonin turnover rate as measured by the serotonin metabolite 5-hydroxyindolacetic acid (5-HIAA) and increases anxiety and consumption of drugs of abuse. Moreover, such monkeys show a lack of reciprocal behavior, appear to feel threatened easily, and react with increased aggressive acts, all directly correlated with reduced levels of serotonin turnover and altered transporter availabilities in the brain stem (Heinz et al., 1998a).

Altogether, acute increases in serotonergic neurotransmission appear to promote positive emotions, while chronic depletion has been associated with increased levels of anxiety, social withdrawal, and, in some cases, impulsive aggression. Accordingly, a substantial number of studies have focused on the role of serotonin in impulsivity.

There is indeed a vast literature on the topic of serotonin, impulsivity, and aggression; however, as a caveat, one has to note that definitions of impulsivity vary considerably between such studies. For example, in some studies, impulsivity is defined as delay discounting (i.e., preferring a small immediate reward instead of a larger reward that is only available later). Other definitions of impulsivity focus on a reduced ability to stop ongoing movements, reductions in attention, unsolicited aggressive acts, or even the mere presence of criminal behavior (Heinz et al., 2001). Varying definitions would not be a problem if the different constructs are tightly correlated when measured with a standardized test; however, this is usually not the case, and the different aspects or facets of impulsivity are often not even significantly correlated with each other in spite of all being relevant (e.g., with respect to relapse prediction in alcoholism) (de Wit, 2008; Rupp et al., 2016). Moreover, some of the constructs meant to measure impulsivity appear to be highly dependent upon social status; for example, depending on how poor you are and how much in need of immediate monetary gains, you may prefer a low but immediately available amount of money over a larger but delayed sum; accordingly, social status was more strongly correlated with impulsivity than any serotonergic state or trait marker (Heinz et al., 2002).

Therefore, in order to better understand serotonergic effects on mood states and reactions to reward and punishment, serotonin effects on different brain areas during processing of aversive and appetitive cues as well as their association with computational models of decision making have to be taken into account. While serotonergic neurons originate in the brain stem, they innervate a wide array of brain areas including the amygdala and further limbic brain regions associated with emotion processing and conditioning (figure 5.4) (Büchel and Dolan, 2000).

Accordingly, genetic variances in serotonin transporter availability have been found to modulate amygdala activation elicited by aversive cues, although meta-analyses have shown that the effect is rather weak and was even absent in some studies (Hariri et al., 2002; Heinz et al., 2005a; Munafo et al., 2008). One problem with assessing amygdala reactivity to aversive cues is the choice of baseline: several studies used a defined baseline (e.g., an affectively neutral stimulus), while other studies just used undefined states (e.g., when watching a fixation cross and waiting for new aversive or positive cues); the latter situation in a dark and narrow space such as a brain scanner appears to be aversive for some subjects and indeed was associated with elevated amygdala activation compared to neutral stimuli among persons carrying a serotonin transporter genotype associated with increased stress vulnerability (Heinz et al., 2007).

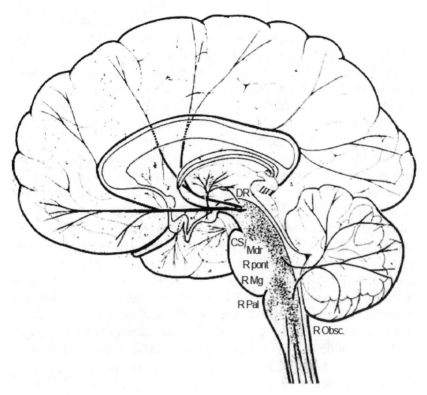

Figure 5.4
The serotonin system originates in the raphe nuclei and projects to an abundance
of cortical and subcortical brain areas including the amygdala and frontal cortex.
Source: Modified according to Heinz et al. (2012b).

Beyond serotonin, dopamine appears to play a role in modifying amygdala activation elicited by aversive but not positive cues (Kienast et al.,
2008). However, one should not identify amygdala activation with the
neurobiological correlate of negative affect—complex phenomena such as
affective states appear to be associated with an interaction between different
brain areas, notably the amygdala and brain regions implicated in emotion
regulation and self-related processing of information including the cingulate and prefrontal cortex (Heinz et al., 2005a; Kienast et al., 2008). Negative mood states were high in healthy volunteers only if such interactions
between the amygdala and cingulate cortex were weak (Kienast et al., 2008).
Likewise, a dimensional approach assessing several mental disorders did

not find evidence for elevated amygdala activation in subjects with major depression or anxiety, neither was amygdala activation directly associated with the severity of negative mood states (Hägele et al., 2016).

Unfortunately, there is no computational model with respect to serotonin function that has so widely been accepted like the account of phasic dopamine encoding of prediction errors suggested by Schultz, Dayan, and Montague (1997). There are even disputes whether the overall effect of serotonergic neurotransmission is affectively positive or negative—while the evidence discussed above with respect to the effects of drugs of abuse such as ecstasy on the one hand and of chronic serotonin depletion on the other appear to support the notion that chronic reduction in serotonergic neurotransmission promotes feelings of anxiety and insecurity, others rather suggest that serotonin acts as a substance encoding aversive outcomes (Heinz et al., 2001; Meyer, 2013). Altogether, effects of drugs of abuse and results of different animal studies suggest that tonically elevated serotonin levels promote feelings of security, while longer-lasting reductions in serotonin neurotransmission have been associated with increased processing of threatening environmental stimuli and—depending on further individual factors including prefrontal serotonergic neurotransmission and its effect on different receptor subtypes—a fight-or-flight reaction with respect to such threatening cues (i.e., anxious withdrawal or aggressive reactions) (Heinz et al., 2011). In this context, it appears to be important to differentiate between the roles of serotonergic neurotransmission in different brain areas. In the amygdala, serotonin dysfunction appears to promote increased activation and hence encoding of aversive cues, while in the frontal cortex, serotonin depletion has been associated with impairments in task-adequate switching of choices (Clarke et al., 2004; Heinz et al., 2011). For example, serotonin depletion in the orbitofrontal cortex disrupted probabilistic discrimination learning and decreased the ability to suppress responses (Rygula et al., 2015). In the same visual discrimination task, serotonin depletion in the amygdala impaired response choice by increasing the effect of false (misleading) punishment and reward in this task (Rygula et al., 2015). Computational modeling of behavior showed that the sensitivity to reinforcement was affected, suggesting a role of serotonin in learning driven by negative and positive motivational outcomes. Furthermore, cognitive inflexibility was increased after prefrontal serotonin depletion, with increased perseverative responses to previously rewarded stimuli in non-human primates after selective serotonin depletion in this brain area (Clarke et al., 2004). Finally, tryptophan depletion, resulting in reduced availability of serotonin, impaired the inhibition of responses to punishing outcomes (Robinson, Cools, and Sahakian, 2012). The authors suggested that serotonergic neurotransmission may contribute to resilience (i.e., it reduces the prediction of punishment), while lack of serotonin after tryptophan depletion disinhibits punishment-associated responses.

Altogether, these findings point to complex interactions between serotonin and computation of rewards and punishments. Serotonin depletion promotes negative mood states hypothetically via altering sensitivity to the anticipation of punishment and its behavioral effects.

6 Evolutionary Concepts of Mental Disorders— A Unifying Framework?

So far, our discussion of basic learning mechanisms and their neurobiological correlates in mental functions and dysfunctions has left us with at least two dimensions: first, a dimension of a positive affect and learning from reinforcement, which is associated with dopamine neurotransmission in the ventral striatum and includes attribution of salience to reward predicting, Pavlovian conditioned stimuli, as well as habitual versus (more) goal-directed ways of decision making; second, a negative affect dimension associated with processing of aversive and threatening cues and involving serotonergic modulation of the amygdala and further limbic brain areas. However, our review has already shown that neurotransmitter systems are not limited to one aspect of either positive or negative affect but rather, like dopamine in the amygdala, can also contribute to encoding aversive events and stimuli (Kienast et al., 2008). Moreover, core areas of the so-called brain reward system including the ventral striatum or so-called fear circuits like the amygdala should not be considered in isolation, but rather understood as core areas of information processing embedded in wider networks that include sensory input as well as subcortical and cortical stimulatory and inhibitory feedback. For example, different parts of the dorsal and ventral striatum have been associated with motor and motivational aspects of active behavior (Alexander, DeLong, and Strick, 1986), while the amygdala and its subnuclei have been associated with different networks contributing to emotion regulation, Pavlovian conditioning, as well as modification of the activity of the autonomous nervous system regulating, for example, breathing frequency or heart rate (Robbins and Everitt, 1999; Parkinson, Robbins, and Everitt, 2000). Alterations of reward circuitry have been implicated in addiction (with drugs of abuse strongly stimulating dopamine release), in schizophrenia and other psychotic episodes (which will be discussed later), and in major depression, where dysfunction of the reward system may contribute to anhedonia or impaired motivation (Heinz et al., 1994; Di Chiara and Bassareo, 2007; Hägele et al., 2015). We will discuss such categories of mental disorders (affective disorders, psychotic disorders, addictive disorders, etc.) in the next chapters and assess how far a dimensional approach that focuses on learning mechanisms is able to explain the key symptoms of these disorders. Before we do so, however, we have to discuss another dimensional approach to mental disorders—an evolutionary account of mental functions and dysfunctions as proposed by John Hughlings Jackson (1884), which has influenced biological as well as psychotherapeutic and psychoanalytic accounts of mental illness (Heinz, 1998; Heinz, 2002b).

Jackson's influence on models of mental disorders can hardly be overestimated. Influenced by Charles R. Darwin, Jackson suggested that human beings are not perfectly created by God, as previously believed according to religious accounts of human development, but rather are characterized by an evolutionary development; and so is the human brain (Jackson, 1884). Previous accounts of mental disorders mainly assumed that a perfectly created mankind suffers from "degeneration" in various degrees, and that different mental disorders are expressions of the severity of this degenerative process. For example, it was assumed that less severe forms of degeneration can manifest as nervousness or excessive and unconstrained desires, while suicidal ideation, cognitive impairments, and malformations characterize increasingly severe levels of degeneration (figure 6.1) (Morel, 1857).

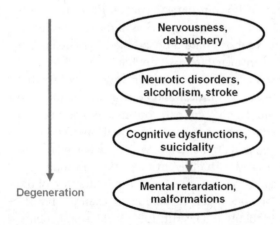

Figure 6.1
Degeneration theory according to Morel (1857). Degeneration theory suggests that mankind was created perfectly but suffers from a uniform degenerative process, which manifests in different yet increasingly severe forms of mental and neurologic symptoms and impairments (Morel, 1857).

Degenerative accounts of mental disorders were apparently inspired by medical experiences with syphilis: a "lapse of morals" (e.g., sleeping with a prostitute and getting infected), already (mis-)interpreted as a potential sign of "degenerative" character traits, can lead to severe malformation in the offspring. One has to remember that circa 1900, about a third of all patients in major psychiatric hospitals in Europe suffered from syphilis, with the bacterium not yet having been detected. Accordingly, there was considerable ignorance regarding early and late manifestations of this infection and a considerable lack of knowledge about the fact that malformations in children born in families who experienced early stages of syphilis are due to intrauterine infections with *Treponema pallidum* and not caused by some miraculous degenerative process that is promoted by moral failures and punitively creates new and ever more severe forms of mental and somatic impairments in each consecutive generation.

As important as a fear of degenerative illness was at the end of the nineteenth and beginning of the twentieth centuries, it did not fit well with the new evolutionary ideas of how mankind was created and—rather than characterized by degeneration—acquired higher and higher levels of cognitive control and perfection compared to non-human primates. It was John Hughlings Jackson (1884) who solved this apparent paradox and suggested that degeneration plays a role in mental behavior, but, it is limited to disease states, in which evolution is reversed (figure 6.2). Jackson's own term for this reversal of evolutionary development was "dissolution" (Jackson, 1884); other authors such as Sigmund Freud, who quoted Jackson and were inspired by his work, used terms such as "regression" or just stuck with the traditional term "degeneration" (Heinz, 1998; Heinz, 2002b).

Unlike Jackson, Freud (1911a, 1911b) did not insist that at the beginning of the twentieth century, evolutionary theories could already identify the respective contribution of lower versus higher brain centers (i.e., older and phylogenetically supposedly more primitive versus more complex brain areas) to symptoms of mental disorders. Instead, Freud suggested a hierarchy of mental functions, with "primary processes" or wishful thinking as the original and rather primitive starting point and more complex "secondary processes," which require inhibition of drives and coping with outside realities, being developed later. Freud (1911b) assumed that evolutionarily more primitive levels of functional organization appear both in early stages of the phylogenetic development of mankind as well as ontogenetically early in the life of the individual, because Haeckel (1866) had suggested that individual development (ontogenesis) is a brief recapitulation of the evolution of the species (phylogenesis). Indeed, Haeckel's claims are still

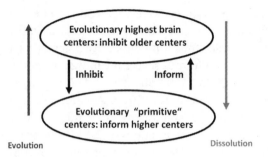

Figure 6.2
The evolutionary model of the interaction of brain centers according to John Hughlings Jackson (1884). Evolutionarily older and more "primitive" brain areas are always assumed to inform (but never inhibit) higher brain centers, while higher brain
centers are always supposed to inhibit lower centers; therefore, dysfunction of higher
centers will always result in disinhibition (and never in inhibition) of lower brain centers. This assumed pathogenic process was called "dissolution" by Jackson, who thus
integrated older theories about degeneration into an evolutionary understanding of
brain development.

valid today with respect to intrauterine development; however, they are no
longer being upheld with respect to postnatal development (Czihak, 1981).
Nevertheless, at the turn of the twentieth century, comparing ontogenesis
and phylogenesis sparked a series of ideas that tried to compare different
stages of human phylogenetic development with different stages of childhood. Degeneration theories had already assumed that there are substantial
differences between human populations, with so-called Caucasians as the
imagined ancestors of Europeans supposedly being the least degenerated,
while Africans were assumed to show the strongest signs of degeneration
(figure 6.3) (Blumenbach, 1795; for a criticism of the "race" concept, see
Heinz et al., 2014).

Identifying phylogenesis with ontogenesis thus offered the dangerous
possibility to compare mental disorders—which were interpreted as resulting from "dissolution," "regression," or "degeneration"—with the behavior
of children and supposedly "primitive men" living in European colonies
around the world (Heinz, 1998). The problem with such phylogenetic com-
parisons was that the phylogenetic ancestors of modern Europeans living
in the "Stone Age" could not be observed, and instead colonialized people
were chosen as supposed exemplars of a "primitive" mentality, thus confirming colonial hierarchies in scientific theory (Heinz, 1998). Nevertheless, at the turn of the nineteenth century to the twentieth century, most

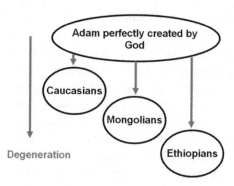

Figure 6.3
Degeneration theory according to Blumenbach (1795). This German anthropologist suggested that different human populations, which he called "races," do not differ in their intrinsic value but with respect to beauty, with what he called "Caucasians" supposedly being the least degenerated and more beautiful than populations living in Asia or Africa. The race concept has been criticized because it is biologically misleading and socially harmful (American Association of Physical Anthropologists, 1996): it wrongly assumes the existence of categorical differences in genetic variance between "races," when in fact there are only gradual differences in allele frequencies among various populations (Livingstone, 1993; Tishkoff et al., 1996). Genetic variation is largest in Africans and tends to decline with human migration out of Africa (Tishkoff et al., 1996; Enoch et al., 2006). Accordingly, humans share the same genes, and only the distribution of specific allele frequencies varies locally without categorical boundaries between populations (Lewontin, 1972; Enoch et al., 2006).

scientists were convinced that Europeans indeed represent the top level of human development, be it with respect to nature or culture, and that all other populations around the world have to repeat exactly the same developmental steps. This belief is reflected in Jackson's unilinear developmental theory of the brain: Jackson did not assume that the brain is a complex interactive network of different brain areas, but rather suggested that it is governed by a strict hierarchy, with phylogenetically older and supposedly more primitive brain centers always informing the respective higher brain area and the respective higher center directly inhibiting the activity of the next older brain region. The resulting model resembles a ladder with successive steps rather than a complex network or web of interactive brain areas.

In Jackson's view, a disease-related loss of a higher brain center would create two problems: lack of activity of the higher area itself, causing "negative symptoms" (i.e., lack of function), and a disinhibition of previously regulated, more primitive brain areas, which are now overactive and create inadequate consequences (so-called positive symptoms). Jackson compared his disease model with a loss of the British government: one would lose the most capable man of society (negative symptoms) and thus have to experience anarchy of the deregulated population (positive symptoms) (Jackson, 1884). Jackson's model thus resembles a military hierarchy, with lower ranks reporting events to higher ranks and higher ranks controlling every movement of the lower military positions. In order to make his model work, Jackson assumed that the phylogenetically youngest and most complex brain areas (including the frontal cortex) are *always* most vulnerable and will lose their function first. The advantage of this rather straightforward model is that it is able to explain all mental disorders with increasing severity as a consequence of more and more of the higher brain areas becoming dysfunctional and successively more primitive brain areas being disinhibited. Such top-down models of central nervous regulation can be very valid in explaining, for example, consequences of a stroke: dysfunction of the first motor neuron descending from the frontal cortex through the pyramidal tract to the second motor neuron in the spinal cord is associated with paresis, that is, lack of function (a negative symptom), while disinhibition of the second motor neuron in the spinal cord is associated with an increase in spastic motor tone (a positive symptom). However, already this example is biologically too simple: spasticity appears to arise from a complex interplay of pyramidal and extrapyramidal motor regulatory centers rather than simply a lack of function of the first motor neuron and a disinhibition of the second.

In spite of these caveats, Jackson's model became widely popular, most likely because unilinear developmental models of the brain as well as of cultural social development dominated scientific discourse in the early twentieth century. The history of colonialized people was widely denied by European scientists, and instead they were seen as somehow miraculously surviving primitive "precursors" of modern mankind. Freud (1913) was one of the few psychiatrists and psychotherapists who reflected upon contemporary prejudices. Nevertheless, he assumed that by studying supposedly "primitive" Aborigines in Australia he could understand psychotic experiences, because Aborigines like schizophrenia patients were supposed to be characterized by magical ideation as an expression of a primitive mind-set (Freud, 1913). Freud's ideas were highly influential for the development of schizophrenia theory, as we will discuss in the next chapter with respect to the theories of Eugen Bleuler and others.

Neurologists such as John Hughlings Jackson and psychiatrists such as Sigmund Freud who focused on biological aspects of mental disorders thus shared a common set of beliefs, including the assumption of a unilinear human development and its reversal in mental disorders, which would result in something like a continuum ranging from primitive to complex mind-sets with respect to evolutionary development and from the irrational behavior of the mentally ill to the rational and goal-directed behavior of healthy European men (with women variably being considered more or less "primitive" and irrational; cf. Heinz, 2002b).

Conceptualizing mental disorders along one single dimension did, however, contradict Kraepelin's attempt to categorize the most serious mental disorders frequently encountered in psychiatric hospitals around the turn of the twentieth century as affective disorders on the one hand and dementia praecox/schizophrenia on the other (Kraepelin, 1913). Indeed, Kraepelin long resisted the allures of evolutionary accounts but shortly before his death publically supported the validity of such approaches and suggested that mental disorders "among Jews, particularly [migrating to Germany] from the East," would often be characterized by "degenerative" and "hysterical traits" (Kraepelin, 1920). Kraepelin's change of mind may at least partially been inspired by his experiences of social unrest after World War I, resulting in councils of workers and soldiers taking power in Munich for brief periods of time and Kraepelin being particularly upset that Jewish Germans were among the revolutionaries strongly supported by the Munich working class. After the unrest was suppressed and mass executions of workers carried out by the military (Hafner, 1994), Eugen Kahn (1919), one of Kraepelin's coworkers, diagnosed some of the rebel leaders including Ernst Toller as being what he called a "psychopath" with "hysterical" character traits. Kraepelin (1919) tried to explain the involvement of Jewish intellectuals and workers in the socialist rebellion by pointing to the alleged "frequency of psychopathic traits" in this "race" and argued against the "unpleasant internationalism" of the "Jewish people" due to their alleged "loss of national roots [nationale Entwurzelung]." Kraepelin further worried that the "German people" (which obviously in Kraepelin's view did not include German Jews) would be particularly endangered by this supposed "loss of national roots," which would be promoted by "marrying persons belonging to foreign populations [Eheschließung mit Angehörigen fremder Völker]," which in Kraepelin's view obviously included German Jews (Kraepelin, 1921). Ernst Toller (1933/2010), one of the revolutionaries arrested and transported into Kraepelin's clinic for diagnostic assessment after participating in a strike to end World War I, described his encounter with Kraepelin from his own point of view:

The director of the psychiatric clinic is the famous Professor Kräplin [sic], who founded in a Munich beer cellar an association to defeat England.—Sir, he shouts at me when I was brought to him, how do you dare to deny the legitimate claims of Germany, this war must be won, Germany needs new space to live, Belgium and the Baltic provinces, it is due to subjects of your kind that Paris is not yet conquered, you are in the way of victory and peace, the enemy is England. [. . .]—The face of the professor gets all flushed, and with the pathos of a manic speaker at public gatherings he tries to convince me of the necessity of Pan-German politics, I learn that there are two kinds of lunatics, the harmless are kept in locked rooms with bars in front of their windows and are called mad, the dangerous ones point out that hunger helps to discipline a population at war and found associations to defeat England, they are allowed to imprison the harmless ones. (Toller, 1933/2010, p. 87)

Thus being driven to accept evolutionary accounts and emphasize the alleged "degeneration" of Jewish Germans, Kraepelin (1920) nevertheless hesitated to accept Jackson's model of evolution and dissolution completely. Instead, he questioned why positive symptoms should only result from disinhibition of supposedly "lower" brain areas and asked why higher brain centers should not also be disinhibited (Kraepelin, 1920). Kraepelin so questioned Jackson's strict hierarchical model of the brain, which was inspired by Jackson's perception of the role of the British government in keeping the "anarchy" of the population at bay. Maybe impressed by the short "coming to power" of cosmopolitan bohemians, artists, and revolutionaries including Ernst Toller, Erich Mühsam, and Gustav Landauer, Kraepelin allowed for the possibility of higher brain centers to make mistakes and thus to create positive symptoms. Rephrasing Jackson, one could state that Kraepelin, driven by his racist interpretation of social unrest during the breakdown of monarchic rule and the establishment of democracy after World War I, allowed for the government to potentially go crazy and thus create "positive" symptoms of madness. Kraepelin died long before the Nazi party grew strong and finally gained power in 1933; however, one of his supervising doctors, Ernst Rüdin (1939), played a key role in enforcing compulsory sterilization of mentally ill patients in Germany during the Nazi regime. These brief and sketchy remarks are meant to caution that choosing models of mental disorders in psychiatry and psychotherapy can have important political and social consequences. Denigration and discrimination of patients with mental disorders can be promoted by careless use of theoretical constructs, a danger that is also present for our current attempt to recast mental disorders as learning dysfunctions—any model of the mind accordingly has to be questioned with respect to its social implications, which we will attempt to do in the concluding chapter of this book. Before we come to this point, we will carefully assess scientific evidence for top-down and bottom-up regulations as well as interactions between different brain areas in key mental disorders and their functional dimensions, and thus try to avoid settling too early with potentially too simplistic models of mental functions of dubious historical background.

7 Computational Models of Learning Mechanisms in Psychosis

How far do computational approaches carry us when we look at major mental disorders such as schizophrenia and related psychotic disorders? To approach this question, it helps to look at the key symptoms that characterize psychotic experiences in schizophrenia and related disorders and to discuss the degree to which they can be understood with respect to the basic dimensions of learning disorders reported above (chapters 2–4). However, this task may not be as straightforward as it appears to be at first sight. Schizophrenia theory and the related discussion about key psychotic symptoms have considerably changed during the past 100 years of research. Currently, the *Diagnostic and Statistical Manual of Mental Disorders*, 5th edition (DSM-5), and the International Statistical Classification of Diseases and Related Health Problems, 10th revision (ICD-10), differ substantially with respect to key symptoms and traditions that define schizophrenia (World Health Organization, 2011; American Psychiatric Association, 2013). Moreover, there have been major conceptual controversies with respect to this disorder, and some researchers even question the usefulness of the term "schizophrenia" itself (Bock and Heinz, 2016; van Os, 2016).

Psychopathologic Traditions and Controversies About Classification

Historically, Kraepelin's dichotomy (i.e., the distinction between major affective disorders with a cyclic course on the one hand and progressively worsening disorders labeled "dementia praecox" on the other) was based on both a fine-grained description of key symptoms and their longitudinal course (Kraepelin, 1913). Intuitively, this distinction also reflects the culturally deeply embedded notion that we can distinguish between "thinking" and "feeling." If we look at the major blocks describing mental disorders in ICD-10, Kraepelin's dichotomy is still reflected in the distinction between current

Exogenous psychoses (brain organic syndromes)	Endogenous psychoses	Variations
Acute e.g., delirium	The group of schizophrenias ICD-10 block F2	Neuroses (trauma & conflict-related causes)
Chronic e.g., dementia	Major affective disorders (unipolar & bipolar depression) ICD-10 block F3	Personality disorders (traits)

Figure 7.1
Dichotomy of endogenous psychoses in ICD-10 blocks F2 and F3.

ICD-10 blocks F2 (schizophrenia and related psychotic disorders) and F3 (mood [affective] disorders; World Health Organization, 2011) (see figure 7.1).

It was Eugen Bleuler who coined the term "the group of schizophrenias" in 1911 (Bleuler, 1911). Bleuler's work was deeply inspired by his interaction with Sigmund Freud and Carl Gustav Jung (Bleuler, 1906). In accordance with the basic assumptions of John Hughlings Jackson (1884), Freud (1911a) assumed that the evolutionary development of the brain and its mental functions is reversed by every disease process, and that the degree of degeneration or "regression" (in Jackson's terms "dissolution") of this unilinear development determines the severity of a mental disorder.

Given the then popular hypothesis of Ernst Haeckel that "ontogeny recapitulates phylogeny," which today is only considered with respect to intrauterine development (Czihak, 1981), it was assumed that postnatal development reflects species evolution; therefore, pathogenetic reversal of evolution, be it called "regression," "dissolution," or "degeneration" (figure 7.2), was supposed to reveal ontogenetically as well as phylogenetically "primitive" stages of development (Heinz, 2002b).

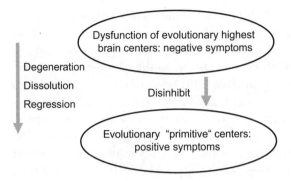

Figure 7.2
Reversing evolution: the concepts of dissolution, degeneration, or regression. According to J. Hughlings Jackson (1884), neurologic as well as psychiatric disorders are characterized by both a dysfunction of the respectively highest, phylogenetically youngest brain centers (resulting in negative symptoms due to the loss of function) and the disinhibition of evolutionarily more primitive centers (manifesting as positive symptoms). Sigmund Freud (1911a) was inspired by this disease model but focused on an account of brain functions rather than centers. In Jackson's as well as in Freud's disease models, the disease process always starts by impairing the highest and most vulnerable center or function and proceeds "downward" in an orderly fashion by reversing the evolutionary development and disinhibiting ever more primitive centers and functions. The severity of any disorder is thus supposed to be characterized by the depth of the alleged reversal ("degeneration," "dissolution," or "regression") of evolution.

Eugen Bleuler acknowledged the multifold phenomena occurring in psychotic disorders and accordingly did not speak of "schizophrenia" as a singular disorder but rather of the "group of schizophrenias" (Bleuler, 1911); nevertheless, he assumed that a core pathogenetic process is at work in all these psychotic spectrum disorders. Inspired by Freud's notion of a deep "regression" of mental functions in psychosis, Bleuler assumed that this inversion of evolutionary complexity has to be profound enough to disinhibit the most basic and "primitive" levels of mental functioning. Sigmund Freud (1911a) assumed that developmental stages associated with adult functioning of the "ego" are reversed in psychotic experiences. Accordingly, patients suffering from paranoia (who are not yet showing complete psychotic disorganization) were supposed to have regressed to a so-called narcissistic level of ego development, in which psychic energy or "libido" is focused on self-representations. Freud postulated that at this level of mental functioning, any interest of the narcissistic ego in other subjects will be limited to persons with phallic attributes (Freud's theory here was focused on male development only). Regressive remanifestation of such an interest of a male person in another male being would then be rejected due to its homosexual connotation and reappear in the distorted form of feeling persecuted by another man (Freud, 1911a)—unfortunately, Freud thus "pathologized" homosexuality in his attempt to explain paranoia. Speculating on the notes of Schreber, a person who wrote a book about his psychotic episodes, Freud further suggested that full-blown psychosis is characterized by an even deeper regression to the ontogenetically most primitive level of "autoerotism," in which the newborn baby is mainly gaining pleasure from sucking its own body parts (e.g., thumb) and is not yet able to experience any caregiver as a person. Instead, Freud (1906) had assumed that the newborn would get dysphoric when being hungry and hallucinate food reward; observing such signs of discomfort, the caregiver (usually the mother) will tend to react and feed the baby, resulting in a coincidence between the "hallucinated" reward (Freud indeed hypothesized such kinds of mental representation in the newborn baby) and the actual feeding process (Freud, 1900, 1911b). Critically, Freud assumed that the newborn neither experiences the caregiver (or her "breast") nor himself as separate persons (Freud, 1938). Bleuler was inspired by this concept of regression into a primitive mental state; however, he was appalled by the erotic connotations of "auto-*erot*ism" and simply cut the erotic part out of the term, thus creating the neologism "autism" (Bleuler, 1911, 1927). In Bleuler's view, this "autoerotic" or "autistic" state of mind is characterized by wishful thinking and hallucination of rewarding objects and not by any attempt to rationally master a person's environment (Bleuler, 1911). While this state of mind was supposed to occur naturally in newborn babies, according to Bleuler it pathologically reappears in schizophrenia due to a dysfunction of higher cognitive processes, which Bleuler imagined to be structured by a realistic and therefore logical representation of reality (Bleuler, 1911). In detail, Bleuler assumed that complex concepts, for example, focusing on who I am or where I live (which he called "complexes"), are logically connected to other complexes, for example, concerning the century in which I was born or the historical events occurring in that time. Bleuler's model of the mind bares similarity to classroom representations of molecules, in which the atoms (corresponding to the complexes) are represented by wooden balls and the connections between the atoms (the logical bridges between the complexes supposedly formed by the experience of reality) are represented by wooden sticks (cf. the depiction of a molecule in Brussels; figure 7.3).

Figure 7.3
A model for cognitive complexes and their associations. The Atomium in Brussels
was designed by the engineer André Waterkeyn and the architects André and Jean
Polak; this image can help to visualize Eugen Bleuler's concept of thought complexes
and the supposedly "logical" bridges formed by the experience of reality that alleg-
edly associates these complexes: the disease process in schizophrenia supposedly de-
stroys these bridging associations, leaving the complexes to be driven and associated
by affective needs rather than reality and logic (Bleuler, 1911).

In Bleuler's concept, the basic pathogenetic process in schizophrenia is
the destruction of the imaginary logical bridges between concepts, which
are then drifting in something like a "sea of emotions" and, according to
emotional needs, can be associated in unusual or even contradictory ways.
According to Bleuler (1911), the key symptom of schizophrenic experience
is the dysfunction of "associations" between complex concepts ("Assoziationsstörung," the impaired association of thoughts), which was supposed
to manifest as "incoherence" in verbal expressions of patients. From this
basic alteration, Bleuler's other key symptoms can easily be developed: if

concepts are no longer linked by logical and empirically based associations but are instead associated or even fused into each other by emotional forces, wishful thinking will prevail over rational mastery of the environment (thus explaining the key symptom of "autism"). Moreover, Bleuler (1911) assumed that in the absence of a realistic and logical order of thought processes, two contradictory complexes can dominate mental functioning at the same time, particularly if they are both linked with strong affect, thus explaining another key concept of Bleuler's understanding of schizophrenia: ambivalence. Finally, cognitive complexes can appear that are not associated with adequate affect, thus explaining "affective flattening" and other disorders of affective responding (Bleuler, 1911).

Bleuler's concept was based on state-of-the-art association experiments of his time. Particularly, subjects were confronted with a certain word (e.g., the word for the color black) and asked to freely associate other words that come to mind. Eugen Bleuler, Carl Gustav Jung, and others had observed that such chains of verbal associations are often interrupted or at least characterized by a substantial prolongation of reaction times whenever patients encountered concepts not only occurring in the current chain of verbal associations but also playing a role in their delusions. For example, a patient who would associate the word "lock" after being prompted by the word "door" may display substantial prolongation of his reaction time in the case where he is convinced to be a persecuted nobleman, because in German the word "lock" and the word "palace" (of the supposedly persecuted nobleman) are similar (both are called "Schloss" in German; Jung, 1907). As experimentally valid as this approach was, it resulted in a clinical criterion for schizophrenia that is quite open to interpretation—incoherence of verbal utterances. Indeed, beyond severe forms of disorganized speech, it can be quite difficult to distinguish everyday problems expressing unusual experiences, alterations of language production during highly affective states, and mild stages of psychotic incoherence of speech. Given that psychiatrists in hospitals usually encountered patients who were quite upset to be admitted (often against their will) to the institution, incoherent argumentation due to anxiety and intimidation may easily have been mistaken for a key symptom of schizophrenia.

In any case, the decision whether a speech is incoherent or not rests with the observer. During the rule of National Socialism in Germany, when patients with schizophrenia were legally subjected to sterilization and illegally killed in high numbers, Kurt Schneider (1942) suggested to base the diagnosis of schizophrenia not on symptoms depending on the "impression of the observer" but rather on symptoms reported by the patients themselves. Accordingly, Schneider suggested that specific descriptions of complex auditory hallucinations ("voices commenting" what a person does or "voices arguing with each other"), specific alterations in the self-ascription of one's own thoughts (e.g., "thought insertion" or "thought withdrawal" by an alien force), and so-called delusional perceptions (i.e., the perception of objects or acts deeply imbued with delusional meaning, which cannot be corrected when confronted with falsifying evidence) should be used as key symptoms to diagnose schizophrenia. Schneider (1942) further suggested that all symptoms not verbally reported by the patient but rather depending on the impression of the observer should be treated as being of only secondary value for the diagnosis of schizophrenia.

Historically, Schneider's approach was tremendously successful and inspired standardized assessment of psychotic disorders using (albeit somewhat variable) lists of his first-rank symptoms outlined above. For example, the international study of the World Health Organization (WHO) that measured prevalence of schizophrenia in Europe, Africa, and Asia was based on assessment tools that strongly relied on first-rank symptoms (Sartorius et al., 1986; Jablensky and Sartorius, 2008). However, in recent years, there has been a shift of focus toward so-called negative symptoms, and questions about the validity and reliability of first-rank symptoms created a considerable dispute (Barch et al., 2013; Soares-Weiser et al., 2015; Heinz et al., 2016a). For example, occurrence of first-rank symptoms in what was supposed to be "pure" affective disorders was reported by Carpenter, Strauss, and Muleh (1973). However, such controversies are mainly due to traditional differences between nosological systems developed by the WHO (ICD) and the American Psychiatric Association (DSM): following Karl Jasper's idea that the most severe mental disorders should successfully be excluded before less severe disorders can be diagnosed, Schneider (1942, 1967) assumed that the presence of first-rank symptoms suffices to diagnose schizophrenia even in the presence of additional symptoms, which can be attributed to affective or neurotic disorders. This approach is reflected in ICD-10 (World Health Organization, 2013), where the presence of first-rank symptoms co-occurring with symptoms of a major affective disorder would automatically require diagnosis of a "schizoaffective disorder," while this is not necessarily the case in *Diagnostic and Statistical Manual of Mental Disorders*, 4th edition (DSM-IV) (American Psychiatric Association, 2000). In fact, in DSM-IV, first-rank symptoms did not suffice to diagnose schizoaffective disorder in case they occur in conjunction with major affective symptoms; instead, sufficient duration and lack of coincidence of psychotic and affective symptoms was of key importance for the diagnosis of schizoaffective disorders. Accordingly, Carpenter was able to claim that first-rank symptoms can occur in affective disorders in the absence of schizophrenia, while following the logic of ICD-10, the same case would have to be diagnosed as schizoaffective disorder, which is accordingly understood as a variant of the schizophrenia spectrum disorders that also includes brief psychotic episodes (ICD block F2). Such controversies show how conceptual differences produce different diagnostic decisions even when confronted with the same phenomena.

In the 1980s, Nancy Andreasen (1982) promoted a considerable interest in negative symptoms including motivational deficits (avolition/apathy), social withdrawal, and impoverished speech (alogia) (see also Andreasen, 1990). Diagnosis of most of these symptoms relies on evaluating the expressive behavior of a patient, hence the impression of the professional, which is why Schneider (1942, 1967) did not trust their clinical reliability and instead focused on evaluating symptoms manifested via the self-reports of his patients. Accordingly, a focus on "negative symptoms" appears to be closer to Eugen Bleuler's concept of schizophrenia (Bleuler, 1911) than to Kurt Schneider's theoretical approach (Schneider, 1942). However, what figures as negative symptoms in current psychiatric classification systems (Andreasen, 1982; Andreasen, 1990) would have mainly been conceptualized as symptoms of disinhibition (and hence as "positive" and not "negative" symptoms) by Eugen Bleuler (1911). Indeed, our current labeling of such symptoms as "negative" is inspired by the fact that they supposedly represent a "loss of function," for example reduced motivation, reduced speech production, or reduced social participation. Bleuler, in contrast, was inspired by Freud's functional understanding of the hierarchical brain model proposed by John Hughlings Jackson, who had assumed that negative symptoms result from the lack of functioning of evolutionarily higher brain centers, while positive symptoms are caused by the disinhibition of more primitive brain areas (Jackson, 1884). Accordingly, Bleuler assumed that the disorder of verbal and hence mental associations ("Assoziationsstörung") is a key pathogenetic process, which interferes with logical thinking (thus resulting in what we may today call executive dysfunction or cognitive impairment, a true "negative" symptom), while wishful thinking (autism), inadequate affect ("Affektstörung"), or ambivalence were supposed to result from the disinhibition of more "primitive brain" mechanisms. Accordingly, Bleuler would distinguish between executive dysfunctions and cognitive impairments, which he would have attributed to a dysfunction of higher brain areas ("negative symptoms"), and wishful ("autistic") thinking and affective disorders, which are supposedly due to the disinhibition of lower brain areas and hence would constitute "positive" symptoms. Classification of anhedonia is a special case, because Bleuler (1911) focused on wishful thinking ("autism"), which in Freund's concept was related autoerotic joy, and this focus on desires and wishes did not fit well with a concept of anhedonia (i.e., lack of joy), which accordingly was introduced much later into the schizophrenia literature (Heinz and Heinze, 1999). In fact, what we today lump together as "negative symptoms" (i.e., cognitive, affective, and motivational deficits) would have been treated by Eugen Bleuler as separate symptoms with distinguishable causes.

Neglecting Bleuler's model of the mind and using the term "negative symptoms" for a variety of affective, motivational, and cognitive impairments has promoted the mingling of a general notion of deficits in schizophrenia with Jackson's key idea that negative symptoms are caused by failure of the evolutionarily youngest and highest brain areas, notably the prefrontal cortex. Accordingly, there are literally hundreds of studies in schizophrenia research that focus on prefrontal cortical functioning and working memory impairments. At the end of the past century, dozens of studies agreed on reporting a hypoactivation of the prefrontal cortex in association with working memory deficits, which miraculously disappeared when brain-imaging techniques shifted from the assessment of regional cerebral blood flow with positron emission tomography (PET) and single photon emission computed tomography (SPECT) to the measurement of functional activation patterns with magnetic resonance imaging (MRI) (Callicott et al., 2003; Heinz, Romero, and Weinberger, 2004a). When avolition, apathy, and anhedonia are considered negative symptoms alongside impairments of working memory and other executive dysfunctions, this lack of distinction indeed promotes a focus on the prefrontal cortex; in contrast, distinguishing between cognitive, affective, and motivational aspects of these so-called negative symptoms promotes looking for distinct neural correlates of anhedonia or apathy in limbic brain areas such as the ventral striatum (Heinz et al., 1994, 2002; Kapur, 2003).

Clinically, diagnosis of the presence or absence of such symptoms can be difficult, particularly when psychomotor slowing due to neuroleptic blockade of dopamine D2 receptors is associated with reduced psychomotor expression of emotions (Heinz et al., 1998b). As discussed earlier, Schneider's first-rank symptoms were selected to avoid a diagnostic bias due to the subjective perception of the patient's affective and psychomotor expressions by a professional, and instead suggested reliance on the reports of patients regarding their sense of authorship and ownership of thoughts and actions. Schneider's approach had strongly influenced diagnostic criteria up to DSM-IV and ICD-10. However, in DSM-5, specific descriptions of complex auditory hallucinations and the reference to certain forms of so-called self-disorders (inserted thoughts, etc.) were abolished; instead, the manifestation of "any kind of delusion or hallucination" is now considered to fulfill a diagnostic criterion for schizophrenia (American Psychiatric Association, 2013). To date, controversies between DSM-5 and more traditional ICD-10 approaches have not been laid to rest. To avoid misdiagnosis of organic brain disorders that tend to manifest with visual rather than verbal hallucinations and to focus on the expressions of patients rather than relying on impressions of the clinical observer, for diagnostic purposes we suggest to rely on acoustic hallucinations, disorders of self-experience, and delusional perceptions that have to be reported by the patient. This way, the focus on the subjective experience of patients is preserved, and their attempts to report their experience are of key importance (Heinz et al., 2016a).

With respect to the diagnosis of delusions, Schneider strongly suggested not to mistake complex belief systems for delusions and to instead focus on "delusional perceptions" (i.e., specific experiences of outer objects or acts that are deeply imbued with a specific meaning for the deluded person). Two aspects have to be distinguished here: one is the attribution of salience (i.e., importance to objects or acts that appear to be irrelevant for other persons but are experienced as deeply meaningful by the deluded patient), and the other is the experience of the deluded patient that this salience is directly relevant for herself or himself (i.e., the aspect of self-reference). Beyond delusional perceptions, there are also delusional ideas, which are supposed to occur spontaneously, and complex delusions, which represent detailed explanatory models of psychotic experiences and hence of the world in which a certain person lives. Schneider's focus on delusional perceptions was driven by practical reasons; namely, to limit the number of false positive diagnoses of schizophrenia at a time when persons were substantially threatened by the legal and illegal consequences of this diagnosis in Nazi Germany. Schneider insisted that delusions are hard to judge, because their complexity often evades empirical testing; moreover, what may sound implausible could still be correct, and to prove his point, Schneider quoted the case of a female patient who worked in a low-paying job yet claimed that she had a love affair with a very high ranking nobleman and was diagnosed with paranoia; as Schneider (1942) emphasized, in fact she indeed had a child from this person. In order to avoid misdiagnosis, Schneider suggested to focus on delusional perceptions, because it can be directly tested whether the object or act imbued with delusional meaning is indeed neutral and truly unrelated to the deluded person or not.

In our view, key symptoms to diagnose schizophrenia are thought insertion and related disorders of self-experience (e.g., thought withdrawal attributed to alien agents), complex acoustic hallucinations, and delusional perceptions, because they are rather reliable criteria of psychotic experiences. Can the learning mechanisms outlined above help to explain such symptoms? To answer this question, we will have to look at the respective symptoms and their as yet only partially elucidated neurobiological correlates in more detail.

Psychotic Experiences and Disease Criteria: An Anthropological Perspective

Before we proceed to answer the question posed in the previous paragraph, we need to answer the question whether key symptoms of psychotic experience fulfill the medical criterion to diagnose a disease. If so, we then have to consider each individual patient and assess whether these symptoms cause personal harm, be it because they are associated with pain or suffering (and hence fulfill the illness criterion) or because they impair activities of daily living relevant for social participation (the sickness criterion): if either of these two latter criteria is fulfilled in addition to the medical disease criterion, the individual has a clinically relevant mental malady (Heinz, 2014). This careful approach is meant to avoid labeling subjects as mentally ill if they report single medically relevant symptoms including hearing voices or experiencing though insertions, but neither suffer from such symptoms nor are impaired in their social interactions.

So, do complex verbal hallucinations, self-disorders, and delusional perceptions indeed represent impairments of functions generally relevant for human life and survival? With respect to pure survival, this is not obviously the case. It is much more plausible to argue that key symptoms of a delirium such as spatial disorientation are directly relevant for human survival independent of specific circumstances and contexts: universally, human beings are required to be able to orient themselves in space and time, and failing to do so can generally threaten human life. On the one hand, hearing voices that demand that the psychotic patient kills herself or himself can also directly threaten human survival. On the other hand, many voices do not communicate contents that are directly threatening to the life of the hallucinating person. Moreover, many patients are able to resist commanding voices and act otherwise. Likewise, a person that experiences many of her thoughts to be inserted by alien forces may be quite unable to act adequately with respect to her goals; however, some patients experience rather frequent thought insertion and yet are able to carry out all activities of daily living that are necessary for individual survival. Finally, delusions appear to be so widespread among human beings and populations that it is hard to argue how they could not be compatible with human survival: rather, widespread delusions can threaten the survival of others; for example, when they assume that a certain alien group is supposedly "inferior" and has to be exterminated. In this respect, Nazi politics appear to be as irrationally resistant to empirical evidence that falsifies key ideological conviction as any delusion can ever be. Nevertheless, it would not be helpful to excuse subjects who promote inhuman political ideologies by attributing their convictions to some pathologic process akin to a psychotic development. We have already emphasized that the current diagnostic criteria as listed in ICD-10 focus on "delusional perceptions" rather than "any kind of delusions" (World Health Organization, 2011), reflecting Kurt Schneider's warning to focus on delusional perceptions in the diagnostic process and to avoid labeling strange complex convictions as delusions (Schneider, 1942). A case in point to illustrate this warning is the example of a German man who had accused his wife, a banker, that she would help her customers to withhold taxes by organizing illegal money transfers to Switzerland. These accusations were embedded in a long and complex description of failings of the current financial system and the political elites. His wife and her friends reported that he was verbally and physically threatening her, and there was indirect evidence that this person sabotaged the car wheels of these friends. The man was judged to be delusional and sent to a forensic psychiatric institution, where in Germany subjects responsible for criminal acts are treated instead of going to jail in case their acts can be explained by a severe mental illness. Years later, it turned out that at the core of the belief system of this person was a true observation—his wife had indeed helped to illegally transfer money to Switzerland (see https://www.theguardian.com /world/2012/nov/28/gustl-mollath-hsv-claims-fraud/). At the core of the supposedly delusional system was thus a series of perceptions and observations that could have been assessed and falsified but obviously were not tested when diagnosing the mental illness. Idiosyncratic beliefs about the state of public affairs and the financial system may be complex or bizarre; in any case, they are usually difficult to falsify, which is the reason why Schneider (1942) suggested to not rely on the diagnosis of any kind of delusion but rather to focus on single perceptions of real-life interactions and acts. Indeed, transfers of money to foreign banks reported by the German man was such a perception that requires to be falsified before it can be interpreted as a sign of psychosis. This example illustrates again why focusing on delusional perception and testing whether the here-described acts are in fact delusional or—even if implausible—true and hence not a sign of a mental illness do appear to be of key importance when diagnosing a psychotic disorder of any kind. Nevertheless, even when carefully focusing on delusional perceptions, again it is not clear why such symptoms should directly threaten the survival of a person who suffers from true delusional experiences and is unable to distance herself from these ideations even when confronted with very plausible evidence to the contrary.

What self-disorders such as inserted thoughts, delusional perceptions, and hallucinations really jeopardize is not so much immediate survival but rather the ability of a certain person to live in a shared world with others, what the Jewish-German philosopher Helmuth Plessner would call a "Mitwelt" (Plessner, 1975). When living with others, it is very important for social interactions to know whether a certain person is motivated by her own ideas and decisions or is just subjected to commanding voices, whether she utters her own views or views experienced as inserted by alien forces, and whether evidence that is sufficient for most other subjects to reject a certain interpretation of a situation or action is completely ignored by a person with delusional perceptions (Heinz, 2014). Likewise, some affective symptoms encountered in psychosis, but also in severe affective disorders such as unipolar depression or bipolar disorder, often do not directly threaten the survival of the person but rather her ability to live with others; for example, if a person is unable to experience pleasure, this anhedonia can severely limit social interactions whenever a joyful event occurs. Classifying key symptoms of what Bleuler (1911) called the "group of schizophrenias" as indicators of a disease is thus based on the assumption that human beings are essentially dependent upon social interactions. This is quite plausible for early development of any person (Mead, 1912). Adult subjects may decide to withdraw from the world and live in "splendid isolation," and such self-chosen social isolation and withdrawal is per se no sign of mental illness. A mental disorder should only be considered once impairments of functions generally required to live with others including the ability to experience joy are fundamentally impaired and in case such generally relevant impairments are indeed individually harmful, be it that they cause suffering or severely limit social participation (Heinz, 2014).

So far, we have always distinguished between signs of a disorder of the self (like thought insertion) and delusions. This line of argument follows a classical distinction in the German psychopathologic tradition, which distinguishes false beliefs about the environment (delusions) from self-experiences (self-disorders such as thought insertion) on the basis of the hypothesis that there is direct access to one's own intentions and experiences from the first-person perspective, while both the professional observer and the patient have equal access to outside events (Frank, 1991). We find these assumptions to be very plausible; however, there is a substantial philosophical debate with respect to the privilege of first- versus second- and third-person perspectives (Shoemaker, 1996; Pauen, 2010). Philosophical anthropologists such as Helmuth Plessner, whose influence on German psychopathology was limited when the Nazis took power in Germany in 1933 and forced him into exile in the Netherlands, suggested that human beings always experience the world from different perspectives. He suggested that both humans and animals act out of a so-called centric positionality, which centers on their embodied position in space and time; accordingly, they experience the environment as directly surrounding themselves. However, humans unlike animals are always able to distance themselves from this centric position and also experience the world from a perspective that Plessner called "eccentric positionality" (Plessner, 1975). Accordingly, we can feel anxious about being on a stage surrounded by an audience that focuses on us when we have to give a presentation and at the same time experience the situation as if looking at us from the outside and feeling proud, anxious, or embarrassed about how this person performs whom we happen to be. Eccentric positionality is also the basis for experiencing the world from the standpoint of others and thus helps to unfold a shared lifeworld ("Mitwelt") among human beings (figure 7.4).

We do not necessarily have to get into the intricacies of Plessner's or any other account of philosophical anthropology. However, Plessner's system helps us to understand how persons can be alienated from their own activities and why experiencing self-disorders may not threaten individual survival but surely impairs living in a shared world with others. When looking at the world from different perspectives ("positions") in our mind, we need to always be able to refer to ourselves as *our selves*, to our body and mind as belonging to us. Indeed, being familiar with oneself and one's own intentions and acts independent of complex reflection has been described as a hallmark of human experience by philosophers including Sartre (1943) and Frank (Frank 1991, 2012; Heinz, Bermpohl, and Frank, 2012a). However, this prereflective self-awareness may be jeopardized in psychotic experiences. Plessner himself had assumed that human beings can start to "oscillate" between the "centric" and "eccentric positionality," thus running the danger of getting lost in one of these positions and losing the ability to experience the other (Plessner, 1975). Therefore, such oscillation describes a state of vulnerability for pathologic processes: according to this line of thought,

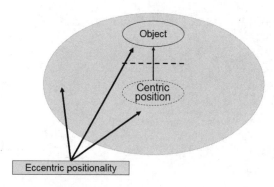

Figure 7.4
Centric and eccentric positionality according to Plessner's philosophical anthropology (Plessner, 1975). All animals and humans experience the surrounding environment from their centric point of view. However, humans also can occupy an eccentric position and experience the world from different perspectives, thus creating a shared world in which we live with others ("Mitwelt"). Based on the work with non-human primates of Wolfgang Köhler (1925), Plessner (1975) suggested that only humans can image an "empty space" and hence remove a box that is located between them and a desired object, and that this capacity is due to eccentric positionality generating an understanding of the constitution of the "world" beyond one's own point of view. Wolfgang Köhler was the only chair of psychology who publically protested the removal of Jewish scientists from German universities during Nazi rule, and Plessner's insistence on the importance of transcending a self-centered position may well have been inspired by rising racism and anti-Semitism during the late 1920s, which attacked every alternative to a supposedly uniform ethnic ("völkisch") point of view.

we can lose our immediate ("prereflective") awareness of our body and mind as belonging to *us* (Heinz, 2014). We can thus get lost in the eccentric position and look at us as if from a distance, without being immediately aware of our own intentions or acts as *our own*. Accordingly, in psychotic episodes many patients describe "inserted thoughts" occurring in their mind or "movements controlled from the outside" enacted by their bodies, which are experienced as alien and attributed to external forces (figure 7.5).

However, when feeling completely encircled by hostile forces, as can be the case in severe paranoia, a subject may lose the ability to distance herself from her current situation, take an "eccentric" position, and reflect upon how others may experience the same situation. A person experiencing severe paranoia can thus be unable to understand that the grimly looking person on the other side of the bus is not threatening her but instead may be sad simply because he lost his job or because of any other occurrence

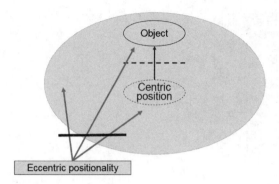

Figure 7.5
Inserted thoughts and other self-disorders as impaired prereflective self-awareness. In some psychotic experience, prereflective awareness of our own mind and body is impaired, with some thoughts experienced as "alien" and "inserted" by external agents or certain movements appearing to be under "alien" control"; subjects may thus get lost in eccentric points of view and fail to automatically identify with their embodied experiences (Heinz, 2014).

completely unrelated to the deluded subject. The deluded subject is thus "arrested" in the "centric positionality," unable to change her perspective and to not feel encircled by a threatening environment (figure 7.6).

Both delusions as well as hallucinations and thought insertions thus appear to result from a loss of flexible prereflective positioning, which is required to simultaneously experience the world from the "centric" and the "eccentric positionality" (Sartre, 1943; Plessner, 1975). Getting lost in either the eccentric or centric position may thus not directly endanger individual survival; however, it severely limits living in a shared world with others, who are constantly challenged to change perspectives when interacting with another human being.

If we follow this line of thought, delusional experiences are characterized by two processes: first, the attribution of salience to occurrences and acts that other persons consider to be irrelevant; and second, by the firm conviction of the deluded person that these signs, events, or acts are centered around herself and are personally highly relevant. Accordingly, attribution of salience to otherwise irrelevant stimuli and experience of personal relevance are the two key aspects that characterize delusional perceptions. Neurobiologically, salience attribution has long been explained by dopamine dysfunction, assuming that stress-dependent or chaotic increases in phasic dopamine release attributes salience to otherwise irrelevant stimuli

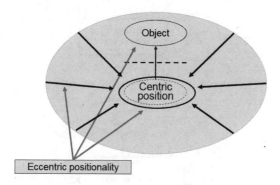

Figure 7.6

Delusions of persecution and impaired transcendence of centric positionality. When experiencing a universally threatening situation as in a delusion of persecution, a person can feel encircled by hostile enemies and is unable to transcend her centric point of view to gain a better understanding of the intentions of others (Heinz, 2014).

(Heinz, 2002a; Kapur, 2003). Salience attribution may also play a role in the development of acoustic hallucinations, particularly when thoughts appear in the "inner poliogue" or personal chain of thoughts that were not expected by the person or are even highly embarrassing for her. If heard aloud, they can be experienced as acoustic hallucinations coming from the outside or inserted into the mind of the person by magical or technical means (Schneider, 1967; Heinz et al., 2012a); when mainly characterized by unexpected content or syntax structure, they can be attributed to an outside force and experienced as "inserted" or otherwise controlled by this alien power (Gallagher, 2000, 2004).

Inserted thoughts and other symptoms of a disorder of self-experience have long been explained as akin to the psychotic experience that one's movements are controlled by an outside force; namely, by pointing to a dysfunction of internal computational mechanisms, which compare the intention to perform a certain act with a forward model of this act (reflecting innervation patterns) on the one hand and sensory feedback elicited by performing the activity on the other (figure 7.7). Discrepancy between intended action and the forward model could then result in impaired authorship or *agency*, while altered sensory feedback (e.g., due to being pushed by another person when carrying out a certain movement) would be experienced as lack of *ownership* of a movement (Frith, 1992). Sensory feedback is supposed to be attenuated when an action is carried out just as intended, and failure to attenuate such feedback may contribute to

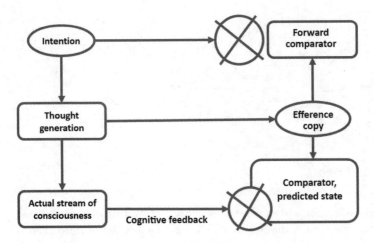

Figure 7.7
A model of self-disorders as impaired intended actions. Movements are supposed to be initiated by encoding a "forward model" of the intended action, which is based on an efference copy of actually descending motor commands; any mismatch with the intended action can be registered as lack of "agency" of the movement; likewise, intentions may be compared with an efference copy of the actually generated "thoughts" and any mismatch may indicate lack of "authorship" ("agency"). With respect to feedback, mismatch between the efference copy and sensory feedback indicates lack of ownership (e.g., my hand may end in an unexpected position because someone bumped into me); likewise, mismatch between forward-modeled thought generation and feedback about the actual stream of consciousness may result in a feeling that I do not "own" certain thoughts (Gallagher, 2004). However, it is quite controversial whether concepts of motor control can actually help to explain cognitive psychotic experiences such as inserted thoughts (Gallagher, 2004; Heinz, 2014). Source: Modified according to Gallagher (2004).

hallucinations. Experimental findings are in accordance with alterations in agency: patients with schizophrenia were impaired in implicitly predicting the outcomes of their actions, which was directly correlated with the severity of hallucinations and delusions; to cope with the task, patients instead excessively relied on sensory feedback (Voss et al., 2010). Impaired self-ascription of movements and associated auditory feedback was also found in prodromal psychotic states (Hauser et al., 2011). However, it is hotly debated whether such models, which plausibly reflect enactment of movements, can be transferred to alterations of thought production, which manifest as the experience of "blocked" or "inserted" thoughts due to the intervention of an alien agent into the person's mind (Gallagher, 2004).

It is quite plausible that we usually feel to be the "author" of our thoughts just like we feel to be the "agent" of our movements. However, while in any motor performance, the intention to act can plausibly be distinguished from the "forward model" that reflects the actual innervation pattern, it is not clear whether an intention to think about a certain topic can really be distinguished from thought generation and the actual chain of thoughts that occurs when we experience an "inner monologue" or rather "polilogue" (due to the many ideas and voices that can enter such an imaginary internal debate). Moreover, with respect to the "ownership" of movements versus thoughts, it can be questioned what specific experience in the process of thinking should be compared to sensory feedback when carrying out a movement (Heinz, 2014). In other words, what alteration in a thought process should cause a person to experience lack of ownership of her thoughts? Thoughts may be heard aloud when subjects think; however, hearing thoughts aloud in one's own mind is usually experienced as acoustic hallucinations, not as inserted thoughts. The specific alteration in inserted thoughts appears to be a lack of authorship; that is, the thoughts are attributed to an outside "agent," rather than to a lack of ownership (i.e., a person experiencing inserted thoughts usually does not claim that she cannot access the alien thoughts that are "strange" due to outside interference). Also, inserted thoughts appear to be out of control of the psychotic subject. However, absence of control over one's own thoughts is per se not limited to psychotic experiences. It also characterizes obsessions, which are experienced as bothersome and unwanted yet not attributed to an outside force in obsessive-compulsive disorder (Bobon, 1983). Altogether, there is currently no "grand unifying theory" with respect to the manifestation of psychotic experiences that explains such seemingly heterogeneous phenomena including commanding or commenting voices, inserted thoughts, or delusional perception with one common mechanism. However, there is substantial agreement in current neurobiological explanations of psychosis, which focus on the interaction of dopaminergic and glutamatergic neurotransmission and point to the role of dopamine in attributing salience to environmental and hypothetically also internal stimuli (Robinson and Berridge, 1993; Heinz and Schlagenhauf, 2010).

Given the heterogeneous symptoms of schizophrenic psychosis (delusions, hallucinations, self-disorders, anhedonia, speech incoherence, etc.), it is understandable that many researchers have searched for a common ground or basic mechanism that explains these multifold phenomena. There is a certain Gestalt intuition that such phenomena belong together and represent a profound alteration in the experience of "reality" (i.e., the way how we live in our immediate environment and in a shared world with other human beings). As discussed earlier, Eugen Bleuler assumed that the fundamental mechanism explaining psychotic experience is a disruption of a basic mechanism structuring cognitions; namely, the association of concepts, which reveals itself as speech incoherence. In Bleuler's theory, this disruption renders cognitions in schizophrenia both illogical and unrealistic. However, when we speak of logic, we focus on the way statements should be structured in order to be meaningful. If a sentence contradicts itself (i.e., this car is black and at the same time is not black), no meaningful message is conveyed. However, phenomena in the outside world, which we tend to call "reality," may not be adequately described with traditional statements, and a sentence like "light is a wave and not a wave" may be illogical but nevertheless refer to problems of describing physical phenomena with concepts and words developed in a different context. When we try to explain unknown phenomena with the language at hand, we necessarily transfer meanings from well-known contexts into unknown territory. In other words, we use metaphors and other figures of speech to chart this unexplored land. For similar reasons, patients who experience psychotic episodes may be at a loss how to describe their experiences using language structures that can appear to be utterly unable to capture the depth and force of psychotic experiences. Wolfgang Blankenburg, another schizophrenia researcher, referred to an expression of one of his patients to describe the fundamental alienation of psychotic patients from established patterns of experiencing and interpreting themselves and their world. His patient described the core of this experience as a loss of common ground, as losing what hitherto was naturally known, acquainted, and unsurprising about the experience of the world (Blankenburg, 1971).

Alienation from previously common and unsurprising experiences also has an affective component; however, it took several decades before it received a prominent place in schizophrenia theory. Specifically, anhedonia, the inability to experience pleasure, was described by Rado (1956) and Meehl (1962) not before the 1950s as a key phenomenon in psychotic experience. The reason for this delay was simply the dominance of evolutionary interpretations of psychotic experiences, in which schizophrenia patients were supposed to be driven by rationally uncontrolled desires and lost in what Bleuler called "autistic" wishful thinking (Bleuler, 1911; Heinz, 1998).

Neurobiological Correlates of Key Psychotic Symptoms in an Anthropological Framework

Key symptoms and mechanisms discussed in schizophrenia research point to different neurobiological correlates. The concept of anhedonia as proposed by Rado (1956) and Meehl (1962) focuses on the loss of the ability to experience pleasure and hence suggests dysfunction of dopaminergic encoding of reward (Wise, 1985). More traditional schizophrenia concepts described a loss of rational control over desires (Bleuler, 1911, 1927), which has traditionally been associated with dysfunction of executive functions associated with the prefrontal cortex (Weinberger, 1996). Finally, Blankenburg's emphasis on psychotic unfamiliarity with common events (Blankenburg, 1971) implicates hippocampal dysfunction and associated alterations in memory, particularly with respect to the encoding and retrieval of well-known versus novel experiences (Lisman and Grace, 2005). In light of the relative lack of gross anatomic alterations in schizophrenia, it has long been assumed that such alterations are either small or that the main problem is altered connectivity between these brain areas (i.e., the prefrontal cortex, hippocampus, and dopamine system ascending from the ventral tegmental area and the substantia nigra to the ventral and dorsal striatum) (Weinberger, 1996).

Given the dominance of top-down theories of mental disorders, it has mostly been assumed that dysfunction of the prefrontal cortex is the primary cause of schizophrenia, corresponding to a Jacksonian impairment of a higher cognitive center, which deregulates dopaminergic neurotransmission, corresponding to the manifestation of phylogenetically old or even "primitive" patterns of experience and behavior, which Jackson would have called "positive symptoms" (Jackson, 1884). Studies in monocytogenetic twins suggested that beyond a substantial genetic contribution, environmental factors such as intrauterine influenza infections or birth complications play a role in the pathogenesis of schizophrenia and related psychosis (Heinz and Weinberger, 2000). Accordingly, animal models were created that tried to capture the interaction between the prefrontal cortex and the subcortical dopamine system. Specifically, early developmental lesions of the prefrontal cortex were studied to see whether they result in a delayed onset of dopamine dysfunction (Pycock, Kerwin, and Carter, 1980). However, in spite of first positive findings, a series of studies failed to show that neonatal lesions of the prefrontal cortex are associated with postpubertal increases in dopaminergic neurotransmission; instead, it was observed that developmentally early temporo-limbic lesions, which impair hippocampal function and associated brain areas or both the ventral hippocampus and the medial prefrontal cortex, interfere with subcortical dopaminergic neurotransmission after puberty (Lipska et al., 1994; Weinberger, 1996; Lipska and Weinberger, 2002). Studies in non-human primates confirmed that developmentally early (neonatal) lesions of the temporo-limbic cortex are associated with altered prefrontal regulations of subcortical dopamine release when the prefrontal cortex is stimulated with amphetamine, which releases dopamine and noradrenaline and thus mimics a stressful situation (Saunders et al., 1998; Heinz et al., 1999). However, anatomic evidence for temporo-limbic alterations in patients with schizophrenia is inconsistent (Weinberger, 1996). Moreover, alterations in brain volume in different cortical areas have been associated with side effects of long-term neuroleptic medication and are hard to distinguish from similar volume reductions present at the onset of psychosis (Ho et al., 2011). Instead, functional imaging studies accessing working memory performance that involve the prefrontal and parietal cortex and that use dynamic causal modeling to look at interactions between brain areas suggest altered connectivity patterns in different subgroups of schizophrenia patients (Deserno et al., 2012). Altogether, our current state of knowledge supports the assumption of altered interactions between key brain areas implicated in executive behavior control (the prefrontal and parietal cortex), memory function, and novelty detection (the hippocampus and entorhinal cortex) and the encoding of prediction errors (the dopamine system).

With respect to dopamine, Grace (1991) distinguished between phasic dopamine release associated with action potentials on the one hand and tonic dopamine levels that are slowly changing in the extracellular space on the other; due to the effects of tonic dopamine levels on autoreceptors, Grace further suggested that schizophrenia is characterized by low tonic and high phasic dopamine release, with high phasic dopamine releases being associated with positive symptoms such as delusions and hallucinations, while low tonic dopamine release should be correlated with negative symptoms such as cognitive or affective impairments. Indeed, tonic elevations in neurologic disorders, for example resulting from overdosing dopamine agonists that bind with high affinity for a long time at dopamine D2 receptors, have been associated with the clinical syndrome of a delirium rather than a schizophrenic psychosis (Heinz et al., 1995a). This means that tonic elevations in dopaminergic neurotransmission were associated with clouding of consciousness, disorientation, and optic hallucinations, all of which rarely occur in schizophrenia (Heinz et al., 1995a). In fact, reduced vigilance or clouding of consciousness is a common symptom of many organic disorders that affect brain function, while schizophrenia patients are usually fully awake and oriented with respect to space and time unless there is a specific distortion due to delusional ideation, which nevertheless lacks the chaotic appearance of disorientation in deliria and other brain organic syndromes (Heinz et al., 1995a, 2016a). Given the profoundly different clinical picture associated with tonic increases in dopamine release, it was quite plausible that Grace (1991) suggested that phasic increases in dopamine firing may characterize schizophrenia. However, our current brain-imaging techniques are too slow to verify or falsify this assumption—at best, we can observe that elevated levels of extracellular dopamine displace radioligands with low affinity from striatal dopamine receptors in psychosis, and indeed studies using amphetamine to release presynaptically stored dopamine as well as studies assessing dopamine synthesis capacity and one study that depleted dopamine and quantified synaptic dopamine concentrations by assessing increased radioligand binding to now dopamine-unoccupied receptors all agreed that there are slight but significant increases in striatal dopamine synthesis capacity and turnover (Laruelle et al., 1996; Abi-Dargham et al., 2000; Kumakura et al., 2007; Howes et al., 2012). However, the timescale of such PET measurement should be kept in mind, as dopamine synthesis capacity and turnover is measured on average for an hour or more, while displacement of radioligands after amphetamine application to assess presynaptic dopamine storage and release is unphysiologic and induces changes in radioligand binding over many minutes, all way above the timescale of phasic dopamine firing patterns in milliseconds as observed in animal experiments (Grace, 1991; Schultz et al., 1997; Schultz, 2007).

A second question is what dopamine actually does when it is phasically released. We have already pointed to the hypothesis of Robinson and Berridge (1993) that phasic alterations in dopamine as induced by the consumption of drugs or the presentation of cues associated with drugs of abuse attribute "incentive salience" to such stimuli and motivate the individual to pay attention to such cues and look for the associated reward. It is highly implausible that one and the same neurotransmitter system is associated with profoundly different clinical symptoms in addiction and schizophrenia. Therefore, we and others (Heinz, 2002a; Kapur, 2003; Heinz and Schlagenhauf, 2010) suggested that phasic alterations in dopaminergic neurotransmission attribute "salience" to otherwise irrelevant environmental stimuli in acute psychotic experiences. As phasic dopamine firing in schizophrenia is assumed to be stress dependent or even chaotic, the clinical picture can differ: while attribution of salience to drug-associated cues should focus attention of addicted subjects to drug and drug-related stimuli, chaotic or stress-associated increases in dopamine firing can attribute salience accidentally to any stimulus or event that is currently occurring. But are there symptoms in schizophrenia that resemble such an accidental attribution of salience to otherwise irrelevant stimuli?

This is indeed the case during early stages of delusion, which are called "delusional mood" (Conrad, 1958). In such stages, one of my patients described that he was "always reading the bible"; however, on this one day when his psychosis started, he suddenly discovered that "on this specific page, there was a message directly for me." He was overwhelmed by the personal meaning of what he read, although he was not yet able to decipher its implications. It took him a while to construct a "rational explanation" of this and similar experiences, resulting in a highly idiosyncratic delusion of sin and grace that centered around this person, who felt overwhelmed by his personal responsibility to save the world from all its evils. Another patient always became psychotic when he met his classmates, who—as he felt—were talking about him. Being in his early twenties, about 10 years earlier he had experienced severe sexual abuse in the boarding school to which he was transferred because of family problems. So it is understandable that being confronted with his classmates, he was afraid that they talked negatively about him; however, this was not what he reported. Instead, he felt that he must have been a prophet and already knew what music titles would be played on the radio 10 years later. What he experienced can be seen as a classical example of attributing salience to otherwise irrelevant stimuli—classmates tend to talk to each other at class anniversary meetings; however, we usually do not assume that all their interactions center around our own person. On the other hand, given the stressful memories of sexual abuse and the observation that stress exposure can increase dopaminergic neurotransmission particularly in the frontal cortex (Abercrombie et al., 1989; Heinz et al., 1999), in the context of a certain vulnerability such experiences may increase phasic cortical and subcortical dopamine firing and hence attribution of salience or "personal meaning" to neutral statements. However, we should emphasize that two processes appear to be involved here: the attribution of salience to outside stimuli and the self-referential aspect of this experience (i.e., the feeling that they directly refer to the psychotic person).

With respect to the attribution of salience to outside stimuli, no study to date directly observed an association between aberrant salience attributions on the one hand and phasic dopaminergic neurotransmission on the other. However, in healthy controls, it has been shown that attribution of salience to irrelevant environmental stimuli is positively correlated with increased dopamine synthesis capacity in the ventral striatum (Boehme et al., 2015), which is significantly increased in schizophrenia patients (Kumakura et al., 2007). Furthermore, attribution of salience to irrelevant task features was high when functional activation of the prefrontal cortex and ventral striatum elicited by reward prediction errors was low (Boehme et al., 2015). In this study, aberrant salience attribution was assessed by testing whether subjects show differences in reaction times to irrelevant features of the so-called salience attribution task. This task, which was developed by Roiser and colleagues (Schmidt and Roiser, 2009), requires the test person to react to stimuli that vary with regard to two features: Animals or nonliving objects are presented either in blue or red. One of these two distinctions (either animals vs. nonliving objects or red vs. blue) is relevant for reward, while the other distinction is not. Aberrant salience attribution is operationalized as showing reaction time differences to the irrelevant (i.e., not rewarded) distinction. If increased dopamine synthesis and turnover is indeed contributing to salience attribution to otherwise irrelevant cues (Heinz, 2002a; Kapur, 2003), then increased dopamine synthesis capacity should not only impair the encoding of reward prediction errors but also contribute to the attribution of salience to irrelevant stimuli and task features, which indeed was the case in healthy controls with high dopamine synthesis capacity in the ventral striatum (Boehme et al., 2015).

With respect to the self-referential aspect of this experience, Pankow and coworkers (2016) observed that aberrant salience attribution measured with the salience attribution task was increased in schizophrenia patients and in subjects with subclinical delusional ideation (although in the latter only to a lesser degree). In parallel, subjects were examined with respect to self-reference (i.e., they had to ascribe trait words either to themselves or to a publically well-known person). Self-ascription of qualities activated the ventral medial prefrontal cortex in healthy controls and individuals with subclinical delusional ideation, while reduced activation of this brain area was found in schizophrenia patients. Notably, and linking impairments in neural encoding of self-referential qualities with aberrant salience attribution in schizophrenia patients, lower levels of ventral medial prefrontal cortex activation elicited by self-referential processing were associated with increased aberrant salience attribution (Pankow et al., 2016). These findings suggest that acute psychotic experiences in schizophrenia are characterized by impaired encoding of self-relevance, which is associated with increased attribution of salience to otherwise irrelevant cues. In other words, the more a certain person is impaired with respect to the self-ascription of characteristic traits, the more she is inclined to attribute importance to environmental stimuli that are irrelevant for most other subjects. Both impairments appear to be associated with altered functioning of the ventral medial prefrontal cortex (Pankow et al., 2016), a brain area that is directly (top-down) associated with the ventral striatum, which has repeatedly been implicated in the pathogenesis of psychosis (Juckel et al., 2006; Schlagenhauf et al., 2014). Correlations are not causations, and such studies do not answer the question whether attribution of salience to irrelevant cues interferes with a person being able to maintain a consistent self-representation or whether, vice versa, alterations in self-representation promote insecurity and an increased focus on environmental stimuli. Hypothetically, both phenomena are two sides of the same coin: increased attribution of salience to some perceptions may render them overly important, while others are neglected. Moreover, at least from the embodied or "centric" position (Plessner, 1975), which perceives the environment centered around the point of view of the subject, there will be only very few highly important events in the environment that are not also personally relevant—only if we distance ourselves from our current position (taking the so-called eccentric position), we can abstract from our current point of view, look at a wider picture, and probably understand that what seems to concern us at the moment is in fact quite unrelated to who we are and what we are currently doing. However, to be able to distance ourselves from our current position, we need to always be able to re-center, to automatically know who we are and where we stand in the world. This means that prereflective self-awareness (Sartre, 1943; Frank, 1991; Heinz et al., 2012a) is required to ensure that we do not get lost in reflections and feel alienated from ourselves, potentially to a degree that renders our thoughts alien enough to appear to be "inserted" by some outside agent. It is quite plausible that the evolution of the representation of environments in animals always places them in the center of action, and it is both computationally costly and potentially dangerous to view the world and ourselves from the outside when we could be attacked by an enemy or approach food or some other reward from our current position. It may be specifically human to be able to distance ourselves from this current position by using language or some other form of abstract symbolization to conceptualize and describe our place in the world. Plessner (1975) insisted that beyond such cognitive ways to de-center oneself, laughing and crying can also help us to distance ourselves from traditional points of view and to provide new perspectives. This human ability to distance ourselves from our current position and to automatically re-center when required may come at a high price—in psychosis, we can start to oscillate between the centric and eccentric positions, at times getting lost in an alienated distance to ourselves or feeling entrenched in the center of personally threatening events.

If so, there can be different reasons why in some psychotic experiences we appear to get stuck in the center of events and feel encircled by a dangerous environment and why we lose our automatic (prereflective) familiarity with our own thoughts and actions and feel alienated: at a biological level, increased salience attribution and hence noise during information processing can render trivial and irrelevant events surprisingly threatening; at a social level, experiences of discrimination and social exclusion may promote feelings of being encircled by a hostile environment and impair the person's ability to automatically identify with her thoughts and actions. We will start by describing biological alterations that may contribute to such experiences and then focus on epidemiologic studies, which describe how social exclusion and racist discrimination may contribute to the manifestation of psychotic episodes.

Toward a Computational Theory of Psychotic Symptoms

Our review of neurobiological studies on dopamine functioning provided evidence that in the ventral striatum, dopamine synthesis capacity is positively correlated with model-based signatures in the prefrontal cortex and negatively correlated with model-free prediction errors in the ventral striatum (Deserno et al., 2015a). Therefore, it is plausible to assume that in schizophrenia, increased dopamine synthesis and turnover in the ventral and central striatum (Abi-Dargham et al., 2000; Kumakura et al., 2007; Howes et al., 2012) results in noisy phasic dopamine release and interferes with encoding of model-based and model-free errors of reward prediction. Indeed, in unmedicated schizophrenia patients, informative errors failed to elicit functional activation of the ventral striatum when participants used a cognitive strategy to infer the current (hidden) state of the decision-making task from the observed feedback (Schlagenhauf et al., 2014). In this context, it is necessary to examine unmedicated schizophrenia patients, because by blocking dopamine D2 receptors, neuroleptics can profoundly interfere with striatal dopaminergic neurotransmission (Gründer et al., 2003). Blunting of reward prediction errors elicited by primary and secondary reinforcers as well as by conditioned reward-predicting cues can interfere with salience attribution and hence motivation to achieve such rewards (Heinz, 2002a). In accordance with this hypothesis, blunting of ventral striatal activation elicited by the temporally surprising appearance of a well-learned reward-predicting cue was associated with negative symptoms such as apathy and anhedonia (Juckel et al., 2006). However, unmedicated schizophrenia patients may well be distracted by hearing voices or focusing on inserted thoughts; therefore, it is important to ensure that patients actually perform the given task following a computational strategy that relies on encoding of errors in the ventral striatum (Stephan et al., 2017). When modeling the performance of individual patients, Schlagenhauf and coworkers (2014) were able to identify one group of unmedicated schizophrenia patients who displayed increased unsuccessful switching in the reversal learning task and hardly performed better than chance. In this group of patients, lack of functional activation of the ventral striatum elicited by informative errors is no evidence for a biological dysfunction but could simply be due to patients not applying a task strategy that requires the computation of reward prediction errors when trying to cope with the decision-making task. A second group of patients performed quite as well as the healthy control subjects, and both this group of patients and the healthy volunteers used informative errors to guide their decision process. Nevertheless, in this group of well-performing patients, informative errors failed to activate the ventral striatum, indicating a neurobiological alteration in striatal encoding of reward prediction errors (Schlagenhauf et al., 2014). This group of patients apparently coped with the task by activating prefrontal cortical areas when computing informative prediction errors, which may help to compensate for ventral striatal dysfunction but at the same time challenge cognitive resources required to guide behavior in complex situations.

Beyond interfering with the encoding of reward prediction errors, noisy phasic dopamine release may also attribute incentive salience to cues that are accidentally present when phasic dopamine discharge occurs (Heinz, 2002a; Kapur, 2003). Direct evidence for this hypothesis is hard to acquire, because human positron emission tomography studies have a quite different time window (minutes to hours) compared to dopamine cell firing (milliseconds). Indirect evidence for dopamine-dependent salience attribution to otherwise neutral stimuli is provided by the above- mentioned study of Boehme et al., which used the salience attribution task (Schmidt and Roiser, 2009) and observed that elevated dopamine synthesis capacity in healthy adults is associated with increased aberrant salience attribution and low ventral striatal encoding of reward prediction errors. Increased salience attribution to task-irrelevant stimuli was also associated with altered encoding of self-referential information in schizophrenia patients in the medial prefrontal cortex (Pankow et al., 2016). Altogether, these studies provide evidence that increased dopaminergic neurotransmission interferes with the encoding of reward prediction errors in schizophrenia patients.

Positron emission tomography studies measuring dopamine synthesis capacity reliably showed elevations in schizophrenia patients (Howes et al., 2012). However, dopamine synthesis is measured for about 1 hour and does not necessarily translate into increased phasic dopamine release. One could also assume that elevated dopamine synthesis increases tonic rather than phasic dopamine discharge (Grace et al., 1991). This is, however, unlikely due to the observation that tonic stimulation of dopamine D2 receptors in neurodegenerative disorders (Parkinson's disease) induced by medication elicits a clinical picture that resembles a delirium with visual hallucinations and clouding of consciousness, while there was no manifestation of key symptoms of schizophrenia such as complex acoustic hallucinations, thought insertions, or delusional perceptions (Heinz et al., 1995a). Moreover, reward prediction error encoding in animal experiments is reliably associated with phasic alterations in dopamine release (Schultz et al., 1997). With respect to localization of dopamine dysfunction in the striatum, functional imaging studies point to alterations in reward prediction error encoding in the ventral striatum, while positron emission studies suggest increased dopamine release in the central striatum and decreases in many cortical brain areas and the midbrain (Kegeles et al., 2010; Slifstein et al., 2015; Weinstein et al., 2017), and dopamine synthesis capacity was elevated in both the ventral and central (associative) striatum (Kumakura et al., 2007; Demjaha et al., 2012; Howes et al., 2012). In summary, these findings suggest that elevated presynaptic dopamine synthesis capacity in the limbic (ventral) and associative (central) striatum interferes with reward prediction error encoding due to a chaotic or stress-induced increase in phasic dopamine release, which attributes salience to otherwise irrelevant, co-occurring stimuli and "drowns" encoding of common primary and secondary reinforcers in noisy dopamine firing (Heinz and Schlagenhauf, 2010).

These considerations suggest a key role for dopamine dysfunction in the ventral and central striatum for the manifestation of both positive and negative symptoms. Attribution of salience to otherwise irrelevant stimuli is implicated in delusional mood, when a multitude of environmental cues appears to be relevant for the individual and appears to contribute to the manifestation of a delusional perception, when an otherwise neutral event or action is experienced as carrying a specific meaning for the deluded person. Moreover, noisy phasic dopamine release may "drown" the encoding of reward prediction errors elicited by common reinforcers and thus contribute to lack of motivation to achieve such rewards (Heinz, 2002a). Robinson and Berridge (1993) suggested that impaired dopaminergic encoding of reward-predicting cues interferes with "wanting" rather than "liking" a reward. In accordance with this hypothesis, blunted functional activation elicited by reward-predicting cues as well as neuroleptic blockade of striatal dopamine D2 receptors were both associated with avolition and apathy (Heinz et al., 1998b; Juckel et al., 2006).

These findings do not necessarily suggest that there is a primary dysfunction of a dopaminergic neurotransmission in schizophrenic psychosis—rather, it is assumed that dopamine dysfunction results from alterations in prefrontal cortical-striatal loops. In accordance with this hypothesis, Kegeles et al. (2000) observed that blockade of glutamatergic N-methyl-D-aspartate (NMDA) receptors in healthy volunteers induces an increase in synaptic dopamine release that is comparable to elevations in schizophrenia patients. As discussed earlier, current neurodevelopmental theories of schizophrenia assume that prefrontal control of subcortical dopamine regulation is impaired because of a developmentally early impairment of temporo-limbic functions, resulting (at least in animal models) in disordered interactions between hippocampal and prefrontal cortical activations and a secondary disinhibition of subcortical dopaminergic neurotransmitters (Heinz et al., 1999; Lipska and Weinberger, 2002). While many studies revealed hippocampal dysfunction in schizophrenia (Heckers, 2001; Heckers and Konradi, 2015), evidence directly linking impaired functional interactions between the hippocampus and prefrontal cortex with subcortical dopaminergic alterations is currently only provided by animal experiments (Lipska et al., 1994; Heinz et al., 1999; Lipska and Weinberger, 2002).

Beyond the specific role of altered dopamine encoding of reward prediction errors in schizophrenia and its potential contribution to delusional mood, delusional perceptions, and motivational deficits, some computational approaches suggest a much wider role for aberrant encoding of prediction errors in schizophrenia (Adams et al., 2013, 2016). These authors suggest in the framework of computational psychiatry that so-called generative models capture the brain's own key functions involved in creating a "model of the world." Following a Bayesian approach (Bayes, 1763), a generative model of the world infers the state of the environment in accordance with prior beliefs, which act as "high-level causes" to generate "low-level data." For example, with respect to perception, differences between prior knowledge and sensory input are computed and constitute a prediction error, which is used to update beliefs in order to create an adequate representation of the environment (Friston et al., 2014; Adams et al., 2016). Bayesian inference thus explains how subjective beliefs change to account for (in this case sensory) evidence. It refers to Bayes's theorem, which states that the probability $[P(A|B)]$ of observing an event A (e.g., a certain belief), given that B (specific sensory data), can be calculated by assessing the probability of observing event B given that A is true ([$P(B|A)$] and by assessing the probabilities of observing events A [$P(A)$] and B [$P(B)$] without regard to each other, is:

$$P(A|B) = P(B|A) \times P(A) / P(B)$$

Bayesian inference can thus help to assess which belief system and hence which "model of the world" best reflects incoming data or objectifiable behavior and to continually update beliefs according to the available sensory data (Stephan et al., 2017).

This rather abstract description can be illustrated with the following clinical example: one of our patients who suffered from a brain tumor in the right hemisphere was not able to see anything present in his left upper field of view. In such cases, it is well known that the brain can by itself "reconstruct" the missing visual information on the basis of knowledge about the surroundings and plausible inferences about what the missing part may look like. For example, when looking at a checkerboard, the "blind spot" will not be perceived and instead the checkerboard is perceived as if it there were no unseen part of the pattern. When asked about such experiences, our patient described an event that had surprised him a lot: "When I recently watched soccer on TV, I was so upset that one player threw a second ball in from the side line. The referee obviously did not take notice and for minutes, the players were continuing to play with two soccer balls. I really got angry until I finally noticed that I must have hallucinated the whole series of events, which occurred in the part of my field of view that is impaired by my tumor." This example, which may be due to additional epileptic activation caused by the brain tumor (Müller et al., 1995), shows how the brain "generates" inferences about the world that directly impact current perception and that are heavily influenced by "priors" (i.e., prior knowledge) about what should be expected when perceiving a certain situation.

On a neurocomputational level, it is assumed that such priors are created at "higher" brain levels and are compared with sensory information provided by "lower" brain levels. Any discrepancy will elicit a prediction error and thus help to adjust prior expectations accordingly (Adams et al., 2013, 2016; Friston et al., 2014). Functional activation of the ventral striatum elicited by errors of reward prediction would thus be one very specific case of prediction error encoding in the brain (den Ouden, Ko, and de Lange, 2012). Indeed, den Ouden et al. (2009) observed prediction errors reflecting learning-dependent surprise in the visual cortex and putamen when carrying out an audiovisual target-detection task. They observed that these prediction errors modulated auditory-to-visual connectivity and that they drive neural plasticity during associative learning. In another task, striatal prediction errors were observed to fine-tune functional coupling in cortical networks during encoding of (failures in) learned predictions about visual stimuli that influence subsequent motor responses (den Ouden et al., 2010). As these studies were carried out in healthy volunteers, specific effects of aberrant prediction error encoding on neural plasticity and associative learning remain to be demonstrated in schizophrenia. However, in light of studies demonstrating altered dopamine function and prediction error encoding in acutely psychotic patients (Abi-Dargham et al., 2000; Kegeles et al., 2010; Schlagenhauf et al., 2014), the studies of den Ouden and coworkers (2009, 2010) suggest that aberrant prediction error encoding directly affects neural plasticity relevant for associative learning.

However, computational theories of psychotic experience to date are often considerably vague with respect to the exact localization and neurobiological correlates of aberrant prediction error encoding in schizophrenia: generally, it is postulated that noisy cortical processing of information can interfere with the precision of high-level "priors," while bottom-up sensory information is largely undisturbed, resulting in increased occurrence of prediction errors due to top-down imprecision (Adams et al., 2013; Friston et al., 2014; for a more complex account of delusion formation emphasizing the impact of ascending loops on circular inference in schizophrenia, see Jardri et al., 2017). It remains to be elucidated exactly between which cortical neural layers or brain areas such top-down and bottom-up processes should occur. Moreover, reduced precision of prior beliefs may result in a rather inconsistent and flexible model of the world, quite in opposition to the rather fixed delusions frequently reported by our patients. A study by Schmack and coworkers (2013) may help to shed light on the role of specific neural networks in delusion information. In this study, ambiguous visual stimuli were presented, which could either be perceived as moving from left to right or vice versa. Movement perception in one specific direction was associated with a respective functional pattern of activation of the visual cortex. Independently assessed severity of delusion ideation was stronger when there was less perceptual stability in the visual cortex, supporting the assumption that noisy information processing contributes to delusion formation. Of key importance is the finding that susceptibility to inducing a false belief (by incorrectly telling subjects that looking through polarizing glasses will remove ambiguity about the direction of stimulus movement) was associated with higher delusional ideation and elevated top-down functional connectivity between frontal areas encoding beliefs and sensory areas encoding perception (Schmack et al., 2013). These observations suggest that noisy information processing in primary sensory areas such as the visual cortex may be due to reduced top-down precision and increased reliance on sensory evidence as suggested by Friston et al. (2014) and Adams et al. (2016); however, such impaired precision of top-down "priors" apparently plays a key role in certain levels of information processing (here: in the visual cortex), while delusion formation is further characterized by the erroneous simplification of complex or ambiguous environmental input, due to strong *a priori* beliefs mediated via top-down effects of the frontal cortex on information processing in other brain regions, which "sculpt perception into conformity with [prior] beliefs" (Schmack et al., 2013).

Comparable to increased fronto-cortical input into brain areas with noisy information processing, increased subcortical dopamine release in psychosis-prone subjects may also represent a compensatory mechanism that is initially aimed at increasing the signal-to-noise ratio in volatile environments. Increased attribution of salience to surprising events may originally help to cope with complex ambiguous information, be the increased noise caused by exposure to unfamiliar stressful environments or primary alterations in hippocampal-prefrontal interactions or a combination of both factors. Neurophysiologic and functional imaging studies indeed found increased noise in cortical processing of information in persons with a genetic risk for schizophrenia and patients with schizophrenia (Winterer et al., 2004, 2006; Rolls et al., 2008). Specifically, a decreased signal-to-noise ratio and an increase in cortical background noise were found in schizophrenia patients and to a lesser degree in their relatives (Winterer et al., 2000, 2004). These findings are in accordance with the general hypothesis of reduced precision in higher levels of information processing in psychosis (Adams et al., 2013, 2016). However, Winterer and colleagues also observed no augmentation of frontal lobe coherence in patients with schizophrenia, which was associated with an increase of noise (Winterer et al., 2000), suggesting that noisy information processing, hypothetically related to temporo-limbic dysfunction, fails to be compensated by altered connectivity between other brain areas including the prefrontal cortex. The observation of Florian Schlagenhauf and coworkers (2014) that reversal learning is best explained by a hidden Markov model (HMM) and that acute schizophrenia patients show impaired ventral striatal encoding of informative errors but a (potentially compensatory) increase in prefrontal activation points in a similar direction: rather than suggesting simple failures of top-down information processing due to a general lack of precision, these observations suggest specific impairments as well as compensatory

activations, which nevertheless come at their own costs including limitations of capacity for other tasks at hand.

Impairments in temporo-limbic processing of information and particularly in novelty detection associated with the hippocampus and ventral striatum have long been suggested to play a key role in the pathogenesis of
schizophrenia (Lisman and Grace, 2005; Lisman et al., 2008; Schott et al.,
2015). While animal models indeed strongly suggest that developmentally
early impairments of temporo-limbic brain areas interfere with prefrontal control of subcortical dopamine release (Weinberger, 1996; Lipska and
Weinberger, 2002), the interaction between these brain areas has not sys-
tematically been examined using computational models in humans. Nevertheless, the abundance of studies pointing to hippocampal alterations in
schizophrenia (Heckers, 2001; Heckers and Konradi, 2015) fits well with the
hypothesis that at the core of psychotic experiences may be a certain lack of
"natural familiarity" with the world (Blankenburg, 1971). In this context,
Lisman and Grace (2005) described how new inputs that differ from prior
information already stored in long-term memory activate a novelty signal
in the hippocampus, which is propagated to dopaminergic neurons in the
midbrain. These midbrain neurons in turn activate projections back to the
hippocampus, which release dopamine and enhance information processing.
Lisman and coworkers (2008) further suggested that in schizophrenia, glutamatergic NMDA receptor dysfunction interferes with the excitation of fast
spiking interneurons and disinhibits pyramidal cells in the hippocampus,
which in turn cause disinhibition of midbrain dopaminergic neurons that
project to the hippocampus, striatum, and further brain areas. The circuit
suggested by Lisman and coworkers (2008) does not include a direct interaction between temporal limbic and prefrontal brain areas in the regulation
of subcortical dopamine release; however, studies in animals that directly
interfered with the function of the hippocampus and adjacent brain areas
confirmed that there are vulnerable developmental periods during which
hippocampal dysfunction affects prefrontal cortical information processing as well as subcortical dopaminergic firing (Saunders et al., 1998; Heinz
et al., 1999), in accordance with the hypothesis that dopamine dysfunction
in schizophrenia is secondary to altered input from afferent systems (Grace,
2016). In this context, early social bonding experiences and their effects on
dopamine and oxytocin modulation of neural networks including the ventral striatum, amygdala, and medial prefrontal cortex remain to be explored
(Atzil et al., 2017).

Altogether, these observations suggest that several processes contribute to psychotic experience: the loss of familiarity with the world,

hypothetically associated with noisy information processing; increased novelty detection mediated by the hippocampus; associated alterations of prefrontal cortical information processing, which have reliably been associated with impairments in working memory and other executive functions; increased top-down effects of prior beliefs mediated by the frontal cortex that may reflect compensatory efforts to cope with an increasingly complex and unfamiliar world; and finally disinhibition of subcortical dopaminergic neurotransmission, which increases salience attribution to otherwise irrelevant stimuli. Furthermore, increased noise of chaotic or stress-dependent dopamine firing can reduce the encoding of errors of reward prediction elicited by primary and secondary reinforcers, thus contributing to a subjective focusing of attention on apparently novel and mysterious environmental cues while reducing attention and motivation elicited by common natural and social stimuli (Heinz, 2002a; Heinz and Schlagenhauf, 2010).

What this model does not explain is the fear and anxiety that many psychotic patients feel when interacting with other subjects, often resulting in delusions of persecution. Here, increased dopamine turnover in the amygdala, which was indeed observed in schizophrenia patients (Kumakura et al., 2007), can directly contribute to increased processing of aversive stimuli and thus bias the experience of the world toward threating cues (Kienast et al., 2008). In accordance with this hypothesis, schizophrenia patients showed increased amygdala activation by aversive stimuli, while the response toward affectively positive cues was blunted (Pankow et al., 2013).

In summary, computational models can help to explain which model-based and model-free decision-making strategies are altered in patients with acute psychosis, and model comparison is required to distinguish between different patient populations that may use different strategies to cope with a certain task (Schlagenhauf et al., 2014; Stephan et al., 2017). Beyond such rather straightforward models of reward learning and decision making in psychosis, complex models are required to simulate computations associated with novelty detection or executive behavioral control as well as the impact of anxiety and other negative mood states (which may be conceptualized as a shift in the conceptual frame of decision making) in psychosis.

Beyond computational modeling, the framework to understand psychosis suggested above is open for a profound reconceptualization of schizophrenia theory: Bleuler (1911) already suggested to speak of a "group of schizophrenias" rather than of a single disease entity. Indeed, there is some evidence that auto-antibodies targeting NMDA receptors may cause psychotic episodes that in some cases resemble acute schizophrenia, while on the psychosocial end of factors contributing to the development of schizo-phrenia, a substantial increase in the incidence and prevalence of psychotic experiences was observed in migrants, particularly when exposed to racist discrimination due to being a visible foreigner and living in a neighbor-hood in which there is apparently little social support (Cantor-Graae and Selten, 2005; Veling, 2013). Indeed, if a key mechanism contributing to the development of psychotic episodes is "unfamiliarity with the world," either biological factors (impaired glutamatergic-GABAergic interactions in the cortex) or social factors (social exclusion and discrimination) may contrib-ute to an apparent increase in "noise" and "complexity" of social interac-tions. A clinical example may help to illustrate this point: When traveling through Mali, our driver told us that he was studying in an East German city when the Communist Party lost its power in 1989. There was an imme-diate increase in racist violence against visible foreigners, who were widely regarded as now unwanted "guests of the former regime," which resulted in violent attacks and even killings of persons with African descent. Our driver informed us that he himself became psychotic: "There were so many killings around me, so many Africans attacked. You never knew who would next turn against you and threaten you. I was afraid and hardly dared to leave my apartment anymore. I started to distrust everyone and could hardly sleep. Finally, I had a severe psychotic breakdown and was admitted to a clinic in Berlin." Our driver told the truth: for an example of racist violence resulting in the murder of an African in this region and time, see the fate of Amadeu Antonia Kiowa (https://de.wikipedia.org/wiki/Amadeu_Antonio_Kiowa). This clinical example shows that psychotic breakdowns can apparently be caused by being confronted with an overwhelming degree of personally threatening information without being able to clearly distinguish between relevant and irrelevant events. In accordance with this hypothesis, it should be noted that psychosis rates in migrants are substantially elevated, while this is not the case in their home country (Heinz, Deserno, and Reining-haus, 2013), suggesting that the increase in psychotic experiences among migrants is promoted by social interactions with the host country. Indeed, psychotic experiences occur more often when "ethnic density is low" (i.e., there are only a few other subjects in the same situation living in your neighborhood), and discrimination as perceived by patients was positively correlated with severity of psychotic symptoms (Veling, 2013). These con-siderations do not contradict the observation that individuals with psy-chotic experiences vary in their vulnerability and resilience, including partly heritable dispositions as well as the degree of exposure to childhood maltreatment (Yuii, Suzuki, and Kurachi, 2007; Grubaugh et al., 2011), but they add a level of social stress factors that include racist discrimination and social exclusion, which can contribute to the manifestation of psychotic experiences (Heinz et al., 2013).

A key aspect in the development of psychotic episodes thus appears to be a lack of a reliable context in which information is interpreted. De-contextualization of information has long been reported in psychotic episodes; however, following the then dominant trend to misread all psychotic experiences as signs of "primitization," the phenomena were mainly discussed under the heading of "concretism." One example is the story of a patient who irritated the director of a psychiatric hospital by always knocking on his door when passing by, because—as the patient explained—there was a sign on the door that read "Please knock." What looks like an act of resistance (we will come back to the subversive aspect of such actions in a moment) turns out to be a misreading of context—the patient indeed just wanted to be polite and did not intend to irritate the director of the hospital or to enact a witty criticism of the abundance of signs with irrelevant orders in hospitals. When interpreting such behavior as "concretism," it is assumed that the patient is unable to understand the abstract concept, in this case the general meaning of the order printed on the sign at the door. However, what the patient actually ignored was the specific context in which this message is relevant—it is only supposed to address those who actually want to enter, not for anyone else. "Concretism," understood as taking statements out of context, thus resembles a well-known strategy to subvert hierarchies by shifting the meaning of orders slightly out of context. *The Good Soldier Švejk* is a famous novel by Jaroslav Hašek and describes a soldier who subverts military hierarchies and senseless drill by taking orders literally and thus subverting their context. We should abstain from labeling similar phenomena in schizophrenia as signs of "primitization" or "concretism": in early stages of a psychotic episode, questioning established beliefs ("priors") and taking statements out of long-established contexts can be witty and creative, but becomes as frightening as psychotic experiences may get when playful shifting of established meaning turns into a nightmare of experiencing an increasingly mysterious and threatening environment. In spite of the aversive consequences of such severe psychotic breakdowns of common beliefs and prereflective familiarity with the world around us, we should not only focus on the limitations and suffering imposed by psychotic experiences. Instead, patients often appreciate certain aspects of their altered states of mind, and relatives, friends, and professionals should respect the rebellious aspects of psychosis, which are often apparent during early stages, before patients feel completely overwhelmed by a chaotic world full of mysterious and threatening information. An example is given by one of our psychotic patients, who, at the end of a clinical visit, when I rose from my chair to say goodbye, told me, "You will rise up high—when you get up from your chair." The patient quite obviously enjoyed his play with words. Likewise, a psychotic patient once cornered the director of a hospital, a well-known biological psychiatrist, by insisting that this doctor should explain the molecular structure of Akineton to him, which is a pharmaceutical drug the patient received to reduce extrapyramidal motor symptoms due to neuroleptic medication. The hospital director quite obviously did not know the molecular structure by heart, which the patient immediately sensed: "Do you know it or don't you know it? Just tell me if you don't, after all, shouldn't a hospital director know the structure of drugs that he prescribes?" When the hospital director had finally retreated into a corner of the room, the patient generously stopped his inquiry and with a big smile told the psychiatrist and the rest of the staff: "The structure of Akineton is pyramidal! This is why it works against extrapyramidal side effects." With a broad smile, the patient left the room, obviously enjoying his play with words. Some weeks later, I met the same patient again, who now was in a deep psychotic crisis, in which he was deeply frightened and disorganized. He could hardly utter a coherent sentence. All he was able to say was: "I need Akineton . . . Pyramids . . . They are good for me." What apparently started as a loosening of semantic context, which allowed the patient to make a witty play with words about pyramids and "extrapyramidal" disorders (which are named after a structure in the brain stem that vaguely resembles a pyramid), completely lost its humorous character when the patient experienced the profound anxieties and cognitive impairments associated with a severe psychotic crisis. Experiencing the lack of precision of higher-level concepts, in this case a clear distinction between pyramids in Egypt and pyramids in the brain, can thus be a curse and blessing at the same time: it allows us to detect the fundamental imprecision of language and the shaky metaphorical ground on which common concepts about ourselves and the world are based, and this experience can lead to a state of exhilaration about the fundamental nonsense of the world, the nonexistence of our assumed securities, and the shallowness of cherished beliefs, but it also confronts us with overwhelming complexity and threatening insecurity and throws us in deep anxious turmoil when confronted with the sheer chaos of being.

Addictions are characterized by strong desires to consume a drug of abuse and by reduced control of drug consumption. Accordingly, drugs are consumed in spite of severe negative consequences for the individual (World Health Organization, 2011; American Psychiatric Association, 2013). Drugs of abuse directly affect brain neurotransmitter systems including dopamine, and all drugs of abuse known to date release dopamine in the ventral striatum or increase dopaminergic neurotransmission in some other way in this brain area. For example, psychostimulants block or even reverse dopamine reuptake, and diazepam inhibits GABAergic interneurons in the brain stem, resulting in indirect disinhibition of dopaminergic neurotransmission (Di Chiara and Imperato, 1988; Lokwan et al., 2000). Drugs of abuse are supposed to enter the brain quickly and to interact strongly with neurotransmitter systems, thus inducing counteradaptive alterations (neuroadaptation) to ensure homeostasis (Koob and Le Moal, 2006). For example, alcohol intake interferes with the function of glutamatergic N-methyl-D-aspartate (NMDA) receptors, thus inducing a counteradaptive up-regulation of these receptors, which balances ethanol-associated inhibition of receptor function and thus contributes to homeostasis; during detoxification, down-regulation of NMDA receptors is delayed and hyperexcitation can occur, which may, for example, clinically manifest as epileptic seizures during withdrawal (figure 8.1) (Tsai, Gastfriend, and Coyle, 1995).

When a subject chronically consumes a drug of abuse, drug effects and counterregulatory neuroadaptations are at a kind of balance (homeostasis); however, in the case of severe alcohol dependence, interruption of alcohol intake during sleep may already suffice to result in withdrawal symptoms that appear once the person wakes up. Such early-morning withdrawal symptoms including tremor and sweating are thus signs of severe alcohol dependence. In fact, neuroadaptation that reduces drug effects manifests clinically as drug tolerance and often motivates the subjects to consume

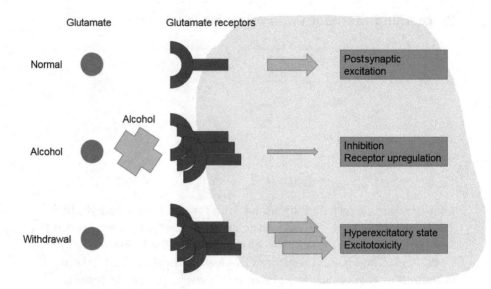

Figure 8.1
Glutamatergic correlates of the development of tolerance and withdrawal symptoms. Tolerance development and withdrawal symptoms can be explained by neuroadaptations that counteract drug effects. For example, alcohol intake inhibits functioning of the glutamatergic NMDA receptor, which is accordingly up-regulated to ensure homeostasis after chronic alcohol intake. Clinically, such counteradaptive regulations contribute to the development of tolerance against direct drug effects and can motivate an individual to increase drug intake to compensate for reduced drug effects. Sudden cessation of alcohol intake disinhibits chronically up-regulated NMDA receptors; their down-regulation takes some time, during which increased receptor function can contribute to hyperexcitation and associated withdrawal symptoms, for example, seizures or (in case of disinhibition of regulatory centers of the autonomous nervous system) tremor and sweating (Tsai et al., 1995; Heinz et al., 2009).

even higher doses of the drug in order to compensate for effects of neuroadaptation. Once drug intake is suddenly interrupted, the previous homeostasis is no longer achieved, and withdrawal symptoms appear that clinically look like the opposite of the drug effect; for example, when alcohol intake or benzodiazepine consumption activates inhibitory $GABA_A$ receptors and induces sedation to some degree, tolerance development is characterized by reduced function of this receptor type (Krystal et al., 2006). During detoxification, drug intake no longer balances the reduction in $GABA_A$ receptors, resulting in a lack of GABAergic function and a disinhibition of certain brain areas such as the frontal cortex, which may result in epileptic seizures.

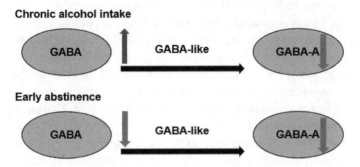

Figure 8.2
GABAergic mechanisms in the development of alcohol tolerance and withdrawal. Alcohol intake interacts with GABAergic neurotransmission (presynaptic GABAergic neurons depicted on the left, postsynaptic cells with $GABA_A$ receptors on the right). Ethanol activates $GABA_A$ receptors and induces a compensatory down-regulation of these inhibitory receptors (or may even exacerbate an already existing deficit in $GABA_A$ function). Once alcohol intake is suddenly stopped during detoxification, there is a delayed recovery of inhibitory $GABA_A$ receptors (or even a persistent deficit in receptor function), which impairs the balance between inhibition and excitation in the brain and contributes to overexcitation during alcohol withdrawal (Krystal et al., 2006; Heinz et al., 2009). Modified according to Heinz et al. (2012b).

Likewise, lack of inhibition of the locus caeruleus in the brain stem may disinhibit noradrenergic neurotransmission and result in the vegetative symptoms often seen during alcohol withdrawal (figure 8.2) (Heinz et al., 2009).

Should We Speak of Drug Addiction or Drug Dependence?

Historically, drug addiction theories have repeatedly shifted their focus. They either tended to be on the more physical side of drug intake, focusing on somatic symptoms during withdrawal and speaking of "drug dependence" (Edwards, 1990; World Health Organization, 2011), or emphasized subjective experiences such as drug craving and reduced control of drug intake in spite of conscious decisions to remain abstinent or to consume less and spoke of "drug addiction" (American Psychiatric Association, 2013).

Behavioral addictions such as a pathologic gambling are also characterized by strong urges to perform the activity in spite of its severe aversive consequences for the individual (e.g., spending a lot of time and losing a massive amount of money when gambling). Addictive non-substance-related behaviors such as pathologic gambling do not have direct pharmacological effects on the brain and hence usually induce neuroadaptation to a lesser degree than that by any consumption of drugs of abuse. Accordingly, withdrawal symptoms are usually milder, and diagnosis of behavioral addictions tends to rely on symptoms such as craving and loss of control (American Psychiatric Association, 2013). Indeed, it was only in 2013 that the American Psychiatric Association decided to include pathologic gambling in the category of addictive disorder. While the *International Classification of Diseases and Related Health Problems*, 10th revision (ICD-10), still uses the concept "dependence" to describe drug addiction and accordingly focuses on withdrawal symptoms as a key indicator of substance-related disorders (Edwards, 1990), the *Diagnostic and Statistical Manual of Mental Disorders*, 5th edition (DSM-5), tends to focus on the motivational aspects of addiction.

With respect to neurobiology, it has been suggested that pathologic gambling is associated with a dysfunction of dopaminergic neurotransmission in the ventral striatum that resembles the changes that can be observed in drug addiction (Reuter et al., 2005). Indeed, directly comparing alcohol dependence and pathologic gambling, there are similarities in alterations of ventral striatal activation elicited by anticipation of reward, while there seem to be differences in processing of loss, hypothetically related to "chasing losses" in pathologic gambling (i.e., the desire to compensate with even higher stakes for previous losses) (Romanczuk-Seiferth et al., 2015).

However, there is one problem associated with focusing on the "addiction" aspect of such disorders: all desires are to some degree characterized by strong urges and reduced behavioral control, be it when wanting to excel in sports, work, or research, being in love, or pursuing a personally highly valued goal. Indeed, Immanuel Kant (1983) suggested that true addictions are characterized by distortions of interpersonal relations. He was not really concerned about people drinking too much, because (at least in his view) they do not interfere too much with their fellow human beings. Rather, Kant suggested that greed and the desire to dominate others ("Habsucht" and "Herrschsucht") are examples of true addictions, because here other persons are no longer respected as human beings with their own value and goals but instead only used as a means to promote egoistic personal goals. Loving another person whom you respect would therefore not fulfill Kant's criterion and should not be labeled an addiction, no matter how strong the respective feelings and desires may be. However, interacting with and caring for other people only when they conform to the rules of a desired gamble or ritual could certainly be labeled as addictive behavior in the

Kantian sense. These considerations suggest not simply using criteria for addictive behavior like independent building blocks that can be piled upon each other to diagnose a disorder; rather, intentions behind actions matter, and normative considerations are implied when we try to distinguish between desires and addictions. Accordingly, the Jewish-German philosopher Helmuth Plessner (2003a, 2003b) warned that in modern democratic societies, we should abstain from deciding whether a certain behavior can be labeled as an addiction by assessing the socially accepted "value" of the behavior. Instead, Plessner demanded that any individually meaningful activity should be respected, no matter how high it is currently valued in society. Indeed, if we want to distinguish between an intense craving to perform research and an intense craving to gamble, we do not need to resort to the current social "value" of gambling versus engaging in research. Instead, it may help to remember Kant's idea that we have to consider whether a certain activity is performed while respecting other human beings as independent individuals with their own values and goals or it instead reduces them to a means to perform the desired activity ("Mittel zum Zweck").

Are Addictions Diseases?

Do the key symptoms of addiction fulfill the key disease criterion; that is, do they reflect impairments of functions relevant for survival and human life? This is certainly true for withdrawal symptoms, which can be directly life-threatening. Tolerance development, as discussed earlier, is the clinical correlate of neuroadaptation to chronic drug intake, which results in a loss of effect of the consumed drugs of abuse and an increased desire or motivation to consume ever higher drug doses to compensate for the reduced drug effect. However, experiencing a strong desire to consume a certain substance or to perform a certain behavior can hardly be called a "life-threatening" activity. Neither does it automatically interfere with the ability to live in a shared world with other human beings—all human life is characterized by acting on our desires and being dedicated to our aims and wishes. However, it can be argued that reducing other human beings to mere figures in a game or objects in a ritual may severely interfere with social participation in a shared lifeworld ("Mitwelt"). Nevertheless, we need to be careful not to pathologize socially unwanted behavior—there is a current tendency to label an abundance of socially unwanted behaviors as addiction, be it "shopping addiction" (you spend too much money and cannot pay it back) or "sex addiction" (too many partners for currently dominant moral values), "Internet addiction" (chatting or even blogging too much), or "work addiction" (you work too hard, will get sick and fail to excel in the future). If we do not want some critical blogger in a dictatorship to be diagnosed with "Internet addiction," we should certainly abstain from labeling any excessive and strongly desired behavior as addictive. We should only speak of an addiction if it is accompanied by tolerance development and withdrawal symptoms or if it disrespectfully reduces other human beings into mere "play toys" for a desired ritual or game. Critical blogging definitely does not fulfill these criteria; neither does dedicated work in the interest of others.

How Can We Become Addicted?

Why do subjects get drawn so deeply into addictive behaviors or the consumption of drugs of abuse? It has been suggested that all drugs of abuse as well as pathologic gambling and other behavioral addictions strongly release dopamine and thus reinforce drug consumption or gambling behavior. This is certainly true for psychostimulants such as amphetamine or cocaine, drugs that release about 10 times the amount of dopamine compared to baseline levels and to the effects of natural reinforcers such as food or sex (Di Chiara and Imperato, 1988; Fuchs, Nagel, and Hauber, 2005; Martel and Fantino, 1996). Behavioral addictions may also release more dopamine than usual, particularly if they are associated with a virtual reality in which "rewards" can occur many times per minute as in some ego shooter games or gambles. Indeed, it has been suggested that gambling is particularly addictive if gains and losses occur at a high speed, and attempts to reduce the "addictiveness" of gambling have often aimed at slowing down the speed of gambling machines. Neurobiologically, it is highly unlikely that phasic dopamine release caused by addictive disorders has effects that are totally unrelated to the effects of phasic dopamine release in schizophrenia and related psychoses. Therefore, we and others suggest that phasic dopamine release attributes "salience" to stimuli associated with the activity or substance that causes the increase in dopaminergic neurotransmission, independent of whether it occurs in addictive or psychotic disorders (Heinz, 2002a; Kapur, 2003). Excessive dopamine release may thus increase the motivational effects of Pavlovian cues that are associated with the availability or consumption of drugs of abuse and render them salient enough to induce active drug seeking (Robinson and Berridge, 1993), while counteradaptive down-regulations in dopamine synthesis, storage, and receptor availability may interfere with salience attribution to non-drug cues (Wrase et al., 2007; Asensio et al., 2010; Volkow et al., 2014). Accordingly, drugs of abuse appear to "hijack" the reward system and bias it toward drug intake.

One frequently cited hypothesis states that drug intake is originally impulsive (i.e., characterized by a desire for immediate reward provided by drug consumption while neglecting long-term negative effects), until it becomes a habit and finally turns out to be compulsive (Everitt and Robbins, 2005, 2016; Belin et al., 2008; Ersche et al., 2016). It has indeed long been assumed that addiction is a kind of deviant behavior that manifests in individuals who do not represent punishment adequately (Eysenck, 1967; Cloninger, 1987a, 1987b). Specifically, it has been suggested that dysfunction of serotonergic modulation of the so-called behavior inhibition system (Gray, 1982; Cloninger, 1987a, 1987b) promotes the manifestation of impulsive actions including drug intake. Studies in rhesus monkeys suggested that serotonin dysfunction as induced by developmentally early social isolation stress is associated with impulsive aggression and increased alcohol intake (Higley et al., 1991, 1992a, 1992b). However, such animals show a series of behavioral impairments, including lack of reciprocity in social interactions, increased anxiety during infancy, misinterpretation of social interactions as threatening, as well as reduced sensitivity to acute alcohol intake, which appears to be due to impaired serotonergic modulation of GABAergic effects of drugs of abuse in the prefrontal cortex (Doudet et al., 1995; Higley et al., 1996a, 1996b; Heinz et al., 1998a).

One has to caution that these findings are largely based on animal models of excessive alcohol intake, which reflect facets of impulsivity (such as seemingly unprovoked aggression that is apparently associated with increased anxiety and impaired social interactions) in monkeys exposed to early social separation stress (Heinz et al., 2001). In humans, the character trait of novelty seeking, which describes a tendency to explore novel environments and to actively seek exciting "sensations" (Cloninger, 1987a), has been identified as one of the factors that are associated with increased alcohol consumption during adolescence (Hinckers et al., 2005). However, such personality traits only explain a small percentage of the variance in alcohol consumption. A factor that has repeatedly been identified as predisposing to excessive alcohol intake is a low sensitivity to acute alcohol effects (Schuckit and Smith, 1996), which has been associated with serotonergic dysfunction after social isolation stress in animal models (Heinz et al., 1998a) as well as with serotonin transporter genotype in adolescence (Hinckers et al., 2006). Apparently, lacking a warning sign that acute alcohol consumption has aversive consequences predisposes adolescents toward excessive alcohol intake. In order to identify such predisposing factors, prospective studies in adolescents are necessary; however, such studies are rare and—due to different regulations with respect to legal drinking age—are largely limited to societies in which it is permitted to assess negative effects of alcohol intake in minors (e.g., the legal drinking age in Germany is 16, while it is 21 in the United States). Further factors associated with excessive alcohol intake are largely limited to the social domain (Whelan et al., 2014).

Beyond alcohol, experimental up-regulation of dopamine D2 receptors in the nucleus accumbens reduced cocaine intake in rats, suggesting that a low level of response to cocaine predisposes to excessive intake (Thanos et al., 2008). In humans, genetic variance in a clock gene contributing to circadian regulation was associated with dopamine D2 receptor availability and cocaine intake, again suggesting that high receptor expression may be a protective factor (Shumay et al., 2012). Notably, nonalcoholic relatives of alcohol-dependent subjects showed elevated dopamine D2 receptor availability compared with healthy controls in the ventral striatum and caudate (Volkow et al., 2006), supporting the hypothesis that reduced availability or sensitivity of dopamine D2 receptors and hence low effects of acute drug consumption may be a factor predisposing toward excessive drug intake.

Low levels of direct drug effects may thus dispose individuals to develop a false feeling of "control" of drug intake. Indeed, one of our patients proudly stated that he could not possibly be alcohol dependent, because he was always the "last man standing" when he and his friends collectively got drunk. This person did not understand that low effects of *acute* alcohol intake do not indicate that the person is also protected from *long-term* neurotoxic alcohol effects. These considerations show why "binge drinking competitions" among adolescents and young adults are dangerous beyond the immediate effects of alcohol on developing brains: individuals who can acutely consume a lot of alcohol are socially rewarded and learn that they are apparently insensitive to aversive alcohol effects, thus drug intake is reinforced and a wrong sense of "immunity" to drug effects is provided.

Beyond low levels of aversive drug effects, individual differences in the experience of positive drug effects may substantially contribute to excessive drug intake. For example, alcohol is known to release endorphins, thus reinforcing drug intake (Gianoulakis et al., 1996). Individual differences in alcohol-associated endorphin release indeed appear to contribute to the risk to develop alcohol use disorders (Gianoulakis, Krishnan, and Thavundayil, 1996; Froehlich et al., 2000). The rewarding effects of endorphins are mediated by mu-opiate receptors, and such effects may be particularly strong in subjects carrying a rare mu-opiate receptor (OPRM1) genotype with high affinity for endorphin (Bart et al., 2005). Accordingly, naltrexone medication, which blocks mu-opiate receptors and thus inhibits the rewarding effects of alcohol-induced endorphin release, may be specifically effective among carriers of this rare genotype (Chamorro et al., 2012; but see Jonas et al., 2014). In vivo, binding of the radioligand [(11)C]carfentanil to mu-opiate receptors competes with endogenous opiate release (Zubieta and Stohler, 2009), and a small number of detoxified alcohol-dependent patients carrying the rare mu-opiate receptor genotype with high endorphin affinity displayed lower ventral striatal carfentanil binding compared to patients with the common genotype (Heinz et al., 2005b). In a prospective study, patients with this rare mu-opiate receptor genotype and low carfentanil binding in the ventral striatum displayed a high relapse risk, again pointing to a role for endogenous opiate release and affinity in mediating treatment outcome (Hermann et al., 2017). However, longitudinal data are required, as exposure to mu-opiate receptor ligands can rapidly induce receptor endocytosis (Roman-Vendrell and Yudowski, 2015), thus potentially explaining conflicting in vitro data (Hermann et al., 2017), which may be due to postmortem alterations in synaptic endorphin concentrations. One longitudinal study in detoxified alcohol-dependent patients carrying the common mu-opiate receptor genotype with low endorphin affinity displayed persistent increases in carfentanil binding in the ventral striatum during the first months of abstinence (Heinz et al., 2005b). However, even such trait-like alterations in alcohol-dependent patients are no proof that they existed before the onset of excessive alcohol intake—they could as well represent a long-term neuroadaptive or neurotoxic effect of chronic alcohol consumption. Indeed, chronic alcohol intake has been associated with neurotoxic effects on diverse neurotransmitter systems including serotonin transporter availability (Heinz et al., 1998c).

With respect to monoamine transporter availability, neurotoxic effects appear to be (at least to some degree) genotype-dependent (Heinz et al., 2000a, 2000b). Furthermore, additional factors such as nicotine consumption or activation of the hypothalamic-pituitary-adrenal (HPA) axis modulate serotonin transporter availability (Reimold et al., 2010; Kobiella et al., 2011). HPA axis alterations have been observed in association with stress exposure (Heim et al., 2008; Flandreau et al., 2012), thus pointing to mechanisms potentially translating environmental risk factors into serotonergic dysfunction. Independent of the sometimes complex causes of reduced serotonin transporter availability in the brain stem, this reduction has repeatedly been associated with negative mood states including anxiety in detoxified alcohol-dependent patients (Heinz et al., 1998c) and major depression (Reimold et al.,

2008). Therefore, chronic alcohol can impair serotonergic neurotransmission and thus contribute to negative mood states, which in a vicious circle can promote further alcohol intake as a maladaptive effort to cope with depression and anxiety.

Altogether, these considerations suggest that both serotonin transporter genotype and alterations in serotonergic neurotransmission after severe social isolation stress can predispose to excessive alcohol intake by reducing its aversive effects, thus rendering subjects more vulnerable to the rewarding effects of drugs of abuse mediated by endorphin and dopamine (Heinz et al., 1998a, 2005b; Hinckers et al., 2006). Beyond low levels of response to alcohol, serotonin dysfunction has also been associated with negative mood states as well as impulsivity, particularly impulsive aggression (Higley et al., 1996a, 1996b; Heinz et al., 2011). Reductions in serotonin turnover (as assessed by measuring the serotonin metabolite 5-hydroxyindoleacetic acid in the cerebrospinal fluid) were indeed correlated with increased aggressive behavior and reduced effects of GABAergic inhibition on frontocortical metabolism (Doudet et al., 2005). In non-human primates as well as in alcohol-dependent patients with high levels of aggressive behavior, low serotonin turnover appears to be associated with both increased levels of anxiety and aggressive behavior (Virkunnen et al., 1994; Heinz et al., 2001). These findings may help to explain the relatively high occurrence of alcohol-induced aggression and violence: acute and chronic alcohol intake interfere with both serotonin function and executive behavior control (Heinz et al., 2011). Alcohol-associated alterations in serotonergic neurotransmission may bias prefrontal and limbic information processing toward interpreting ambiguous environmental stimuli as threatening cues, thus promoting either anxious withdrawal or aggressive attacks, particularly when affect regulation and executive control of behavior is impaired because of acute alcohol intoxication (Heinz et al., 2011). A landmark study in animals with high alcohol-associated aggression observed a blunted prefrontal serotonin release after stimulation of a serotonin receptor subtype (5-HT1B), thus confirming the proposed link between impaired serotoninergic modulation of prefrontal functions and alcohol-associated aggression (Faccimodo, Bannai, Miczek, 2008). However, correlation is not causation, and human studies observing associations between impulsivity and increased drug intake do not answer the question whether increased impulsivity predisposes to risky choices including the consumption of drugs of abuse or whether impulsive behavior results from drug-associated effects on monoaminergic neurotransmitter systems including neurotoxic effects on serotonin transporter availability (Heinz et al., 1998c, 2000a). Also, inter-actions can be multifold, with impulsivity being at least partially promoted by stress-associated impairments in serotonergic neurotransmission and contributing to drug intake, which further impairs monoaminergic systems and thus facilitates impulsive aggression (Heinz et al., 2011).

Notably, several adoption and twin studies provided no evidence for heritability of violent behavior per se, yet indicated that there is a genetic contribution to the combined manifestation of alcohol problems and aggression (Bohman et al., 1982; Bohman, 1996). These findings suggest that while the isolated manifestation of violent behavior is not heritable, there is a genetic contribution to the manifestation of aggressive behavior in the context of alcohol use disorders. In this context, intergenerational violence plays an important role. Children of alcohol-dependent parents are particularly vulnerable to experiencing violence during childhood and adolescence, and exposure to childhood maltreatment increases the risk to display violent behavior, particularly if subjects carry specific alleles inter-acting with monoamine metabolism (Caspi et al., 2002; Heinz et al., 2011).

In summary, with respect to the role of impulsivity in the development of drug addiction, it should be noted that impulsivity is a multifaceted con-struct that includes impaired motor inhibition, discounting of delayed rewards, and impulsive aggression, which have often been found to not be correlated with one another (Rupp et al., 2016). Rather than generally speak-ing of impulsivity, it is therefore preferable to describe the exact mechanism that is supposed to bias a person's behavior toward excessive drug intake. In non-human primates, social isolation stress impairs serotonergic neuro-transmission, reduces the effects of acute alcohol intake, increases threat perception and anxiety, reduces social competence, and increases impul-sive aggression, particularly in male monkeys (Heinz et al., 2011). Animal experiments suggested that increased impulsive aggressiveness as well as a low level of response to alcohol were associated with low serotonin turnover rates and associated alterations in serotonin transporter availability (Higley et al., 1996a, 1996b; Heinz et al., 1998a, 2003b); however, the effects of spe-cific alterations of serotonin reuptake in the brain stem on alcohol sensi-tivity and impulsive aggression may vary considerably (Doudet et al., 1995; Faccidomo et al., 2008). The most consistent factor predisposing to exces-sive drug intake appears to be a low level of response to acute drug effects (Schuckit and Smith, 1996; Hinckers et al., 2006; Thanos et al., 2008; Shu-may et al., 2012). Low sensitivity to acute drug effects is not dangerous per se; however, it may promote drug intake if acute drug "tolerance" is socially rewarded, as in binge drinking competitions. Preventive efforts should thus focus on warning adolescents that they are not protected from harm just because they can acutely tolerate higher drug doses than their friends.

Why Are Addictive Behaviors So Hard to Change?

As discussed above, chronic drug intake can bias the reward system toward drug consumption, and cessation of drug intake can cause aversive with-drawal symptoms that motivate a subject to continue drug use (Goldstein and Volkow, 2011; Heinz et al., 2011). However, even when acute detoxi-fication occurred and patients consciously decided to remain abstinent, many detoxified addicts relapsed when confronted with drug cues or avail-ability of the drug itself (Shaham et al., 2003; Heinz et al., 2009). If indeed one of the most important brain areas associated with dopamine-mediated reward prediction is the ventral striatum, then drug cues such as the sight or smell of the preferred drink should activate the ventral striatum and thus contribute to the motivation for drug seeking and consumption. While this has been observed in some studies (Braus et al., 2001; Grüsser et al., 2004; Wrase et al., 2007; Myrick et al., 2008; Vollstädt-Klein et al., 2010), it has often not been replicated, and a recent review of the literature did not find convincing evidence for consistently elevated ventral striatal activation elicited by drug-related stimuli in alcohol-dependent patients (Huys et al., 2016a). Instead, it was often observed that such drug-associ-ated cues activate the anterior cingulate and medial prefrontal cortex, and that this activation predicts subsequent relapse (Grüsser et al., 2004; Beck et al., 2012; Seo et al., 2015). The anterior cingulate and medial pre-frontal cortex have been associated with a series of functions and activities including attention, error detection, and processing of self-related infor-mation (Carter et al., 1998; Bechara, 2003; Northoff and Bermpohl, 2004; Oliveira, McDonald, and Goodman, 2007). A caveat is that most stud-ies in addicted subjects have been carried out after detoxification, when there is no longer any drug-induced stimulation of dopaminergic neuro-transmission; instead, early abstinence from drugs of abuse is characterized by a slow recovery of counteradaptive down-regulations of dopamine D2 receptors, transporters, synthesis capacity, and release (Volkow et al., 1997, 2015b; Heinz et al., 2004b, 2005c; Martinez et al., 2005, 2012). There are some differences between drugs of abuse: detoxified patients with alcohol, cocaine, and heroin dependence showed blunted dopamine release after psychostimulant application (Martinez et al., 2005, 2012), while detoxi-fied patients with methamphetamine abuse displayed no such alteration in stimulant-induced dopamine release (Volkow et al., 2014, 2015b). As also observed in detoxified alcohol-dependent patients (Laine et al., 1999; Heinz et al., 2011), methamphetamine abusers had lower dopamine D2/D3 receptor and dopamine transporter availability during early detoxification, with only transporter availability recovering in patients who remained abstinent for 9 months (Volkow et al., 2015b). After detoxification, persistent dysfunction of dopaminergic neurotransmission can interfere with the function of fronto-striatal neurocircuits and thus impair motivation and reward-dependent learning mechanisms (Heinz et al., 2004b; Park et al., 2010). As a result, attention can be focused on drug consumption and drug cues, because due to their direct pharmacological effects on the dopamine system, drugs of abuse can activate this otherwise down-regulated "reward system" to a degree that the related activities (drug seeking and intake) are still reinforced. In accordance with this hypothesis, it has been observed that the degree of down-regulation of ventral striatal dopamine D2 receptors is directly associated with increased processing of alcohol cues in the medial prefrontal and anterior cingulate cortex (Heinz et al., 2004b). Impairment of reward-related functions may thus bias addicted individuals to focus their attention on drugs and drug-related stimuli, which are desired exactly because they have strong effects on dopaminergic striatal neurotransmission, even though their repeated consumption in case of relapse will induce further counteradaptive down-regulations in this system, creating a vicious cycle of increased drug intake and decreased drug effects (Koob and Le Moal, 1997).

Addictions and Compulsions

Before we can discuss the effects of neurobiological alterations in drug addiction on reward-related learning and decision making in more detail, we have to further clarify our understanding of addictive behavior and address an important conceptual question: Are addictions compulsions? Addicted subjects often describe a strong urge or "craving" to consume the drug of abuse or to perform the addictive behavior in spite of aversive consequences. Accordingly, it has been suggested that addictions are characterized by "obsessive" thoughts focused on drug consumption and strong motivational states "compelling" a subject to consume the drug of abuse in spite of the conscious decision to remain abstinent (Everitt and Robbins, 2005, 2016). Indeed, persons suffering from obsessive-compulsive disorder (OCD) also report obsessive thoughts, which "force" themselves into their stream of consciousness and—particularly if their content is threatening to the person herself or others—elicit compulsive acts that are often meant to counteract the imagined threat. For example, a person may fear that she is impure and "soiling" others and hence feel compelled to continually wash her hands. However, unlike compulsions in OCD, addictive behaviors are not carried out as often and continually as compulsions in OCD. For example, one of our patients, who suffered from OCD, gambling addiction, and alcohol dependence, described that he always felt compelled to perform certain rituals aimed at preventing his obsessive-aggressive thoughts to harm others, while urges to gamble only appeared around 6:00 p.m., when online gambling was possible, and alcohol cravings were experienced from time to time, particularly when confronted with alcohol-associated cues (Schoofs and Heinz, 2013). Unlike rather permanent obsessions and compulsions, alcohol craving and urges to gamble thus appear to manifest when elicited by specific cues or contexts; for example, when being in a certain place at a certain time where one previously consumed the drug of abuse or carried out the addictive behavior.

In spite of such clinical distinctions, many alcohol researchers, particularly when focusing on animal experiments, suggest that addictive behavior is "compulsive" and characterized by a shift of behavior control from being goal-directed to habitual to finally getting completely out of control and appearing to be compulsive (Everitt and Robbins, 2005, 2016). While it is controversial whether drug craving should really be called an "obsession" and whether habitual drug intake becomes "compulsive" in addiction, there is certainly a shift from goal-directed to habitual drug intake in addicted subjects (Tiffany and Carter, 1998). Indeed, reductions in goal-directed behavior in the two-step task described in previous chapters and other tasks that measure the balance between goal-directed and habitual responding have been observed in subjects suffering from psychostimulant as well as alcohol dependence (Sjoerds et al., 2013; Sebold et al., 2014; Voon et al., 2015; Ersche et al., 2016).

It is quite plausible and occurs often in everyday life that goal-directed actions that are frequently repeated become habits; all training is aimed at such habitization of behavior, be it that we learn how to dance or how to perform a certain sport. Neurobiologically, habit formation appears to be associated with a shift of information processing from the ventral to the dorsal striatum (i.e., from a more limbic and motivationally influenced part of the basal ganglia to areas associated with automatic motor responses) (Haber et al., 2000). However, it is not quite clear how such habits may become compulsions. Tiffany and Carter (1998) suggested that drug consumption often resembles "automatic" behavior patterns (i.e., it is regulated "outside of awareness," stereotyped, effortless, and difficult to control).

Whoever tried to alter a well-established move in sports knows how difficult it can be to "kick the habit"; this even more so in drug addiction, where chronic drug consumption interferes with ventral striatal dopamine function required for flexible goal-directed decision making and biases decision making toward cue-induced habitual drug seeking (Heinz et al., 2004b; Deserno et al., 2015b). However, just because it is difficult to kick a habit, it is not impossible to do so: habits can be changed with practice and time, while compulsions are supposed to be irresistible. Indeed, about 10% to 20% of all alcohol-dependent patients manage to stay abstinent after detoxification even without receiving regular further support (Imber et al., 1976; Helzer et al., 1985). Foddy and Savulescu (2007) even suggested that addictions are "mere pleasure-oriented desires," because unlike true compulsions, addicted subjects can at least momentarily stop drug intake when directly confronted with very negative consequences. For example, when being confronted with death in the so-called gallow's test (subjects will immediately be hung in case they show a certain behavior), a person addicted to a drug will be able to abstain while a person suffering from a true compulsion will not be able to do so. However, Foddy and Savulescu (2007) overlooked the famous statement of Lacan (1986) that a threat to die may deter a subject from engaging in his sexual desires, while it can certainly fail to keep him from committing a murder. Moreover, clinical experience with Tourette's syndrome shows that some complex compulsions can at least be put on hold for short periods of time, while simple tics and compulsive impulses such as coprolalia may manifest even though they are highly embarrassing and the afflicted subject did everything he could to suppress them (Heinz, 1999, 2014). These considerations suggest that rather than constructing a categorical difference between compulsions and habits, they should be seen as parts of a continuum, which ranges from habits that can rather easily be changed to compulsive impulses that can at best be suppressed for a very short period of time. Habitual drug intake in addiction appears to be somewhere in the middle of that continuum—it can be stopped for certain periods of time but may reappear "automatically" when triggered by certain cues, contexts, or stress factors, and it may or may not be associated with a strong conscious craving to consume the drug of abuse (Tiffany, 1999). Nevertheless, habits are hard to change, and addicted subjects are biased toward approach when confronted with drug cues, which was correlated with increased functional activation of the medial prefrontal cortex and ventral striatum (Wiers et al., 2014).

With respect to neurobiological correlates, there are similarities as well as profound differences between OCD and addictive disorders. In light of the fundamental role of fronto-striatal-thalamic circuits regulating human decision making and action planning, it is not surprising that both OCD and addictions have been associated with altered information processing in these neurocircuits. However, findings in the frontal cortex differ between addictive disorders and OCD: Volkow and coworkers (2015a) show that whole-brain glucose metabolism during rest was reduced in heavy drinkers compared to that in controls. Moreover, reduced glucose utilization in the medial prefrontal cortex was associated with increased errors occurring in the Wisconsin Card Sorting Test, which measures executive functions, despite similar mean IQ levels of alcohol-dependent patients and healthy controls (Adams et al., 1993). Likewise in methamphetamine abusers, low levels of orbitofrontal glucose utilization were observed and correlated with reduced dopamine D2 receptor availability in the striatum (Volkow et al., 2001).

In contrast to these findings of reduced frontal glucose utilization in addiction, increased glucose utilization in the orbitofrontal cortex and caudate was observed in in patients with OCD compared to that in healthy controls (Baxter et al., 1987, 1992). After successful treatment with the anti-depressive drug clomipramine, there was a significant decrease in glucose utilization in the right orbitofrontal cortex and caudate nucleus (Saxena et al., 1999). Also, clinical improvement with behavioral therapy correlated with decreases in left orbitofrontal cortex (OFC) glucose utilization (Brody et al., 1998). Increased resting state glucose utilization in OCD has been attributed to a hyperactivation of neurocircuits including the caudate nucleus, which has been implicated in the performance of complex repetitive behavior patterns, and the orbitofrontal, medial prefrontal, and anterior cingulate cortex, which contribute to attention and error detection (Heinz, 1999). Beyond fronto-striatal-thalamic circuits, temporal and parietal brain areas have been implicated in the pathogenesis of OCD (Menzies et al., 2008). Specific obsessions and compulsions appear to be related to different brain areas, with "checking" being associated with activation of dorsal cortical areas, the dorsal striatum and thalamus, "washing" with activation of bilateral ventromedial prefrontal cortex and right caudate, and "hoarding" with the right orbitofrontal cortex and the left precentral gyrus (Mataix-Cols et al., 2004).

With respect to neurotransmitter systems, alterations in fronto-cortical glucose utilization as well as cue-induced brain activation in addiction have been associated with reduced dopamine D2 receptors in the striatum (Volkow et al., 2001; Heinz et al., 2004b), while in the spectrum of OCD, dopamine dysfunction has been implicated in the manifestation of motor tics in Tourette's syndrome (Wolf et al., 1996), whereas compulsive impulses such as coprolalia in Tourette's syndrome as well as compulsions and obsessions in OCD were associated with low serotonin transporter availability in the brain stem (Heinz et al., 1998d; Bandelow et al., 2016).

Altogether, these findings suggest that there are considerable differences between OCD and addiction, with OCD being characterized by elevated and addictions typically associated with reduced glucose utilization in the frontal cortex. Reductions in frontal glucose metabolism in addiction may reflect neurotoxic effects of chronic drug intake, while increases in OCD have been attributed to impairments in error detection and the emotional evaluation of unsuccessful behavior strategies, which contribute to disinhibited information processing in fronto-striatal-thalamic loops (Heinz, 1999; Heinz et al., 2009). With respect to impulsivity, animal studies suggested that motor impulsivity is correlated with reduced cortical thickness of the anterior region of the insula cortex in rats that are also at risk to develop compulsive behavior (Belin-Rauscent et al., 2016), while human studies observed that OCD patients with hoarding symptoms show decreased rather than increased impulsivity and a high sensitivity to punishment (Fullana et al., 2004).

These findings warn us that similarities in reductions of goal-directed behavior in OCD and addiction (Sebold et al., 2014; Voon et al., 2015) can be associated with quite different neurobiological correlates. In line with this hypothesis, blunted ventral striatal activation elicited by reward-predicting cues was found in patients with alcohol dependence, schizophrenia, and (to a lesser degree) major depression (Hägele et al., 2015), but not in OCD patients, who were instead characterized by increased medial and superior frontal cortex activation during anticipation of losses (Kaufmann et al., 2013). Moreover, the medial and superior frontal cortices were less activated during reward anticipation in OCD patients, who also displayed delayed responses for anticipated rewards compared to their responses for anticipated losses, while the reverse was found in healthy participants. These observations do not argue against a dimensional approach to understand mental disorders—indeed, it can be quite useful to group disorders together that are all characterized by reduced goal-directed decision making and a shift toward habits; nevertheless, the construction of these dimensions depends on the level of biological complexity examined in respective studies, with alterations in mechanisms of learning and decision making constituting a middle level that links clinical symptoms with their complex neurobiological underpinnings.

In summary, there are analogies and similarities between obsessions and compulsions in OCD on the one hand and drug cravings as well as reduced control of drug intake on the other; however, most habits are not compulsive, and it is far from clear what are the exact neurocomputational and neurobiological correlates of overwhelming urges to consume a drug of abuse or to reenact pathologic gambling behavior in spite of conscious decisions to abstain.

Why It Can Be So Hard to "Kick the Habit"

If addictions are not simply drug-associated compulsions, why is it so difficult to "kick the habit"? Why is automatic drug intake not stopped, and why are cravings not resisted? There are several possible explanations.

The first one focuses on the strength of the urge to consume the drug of abuse. Overwhelmingly strong cravings may indeed override a conscious decision—you want to lose weight, but your hunger feelings are too strong to resist. Likewise, you may want to remain abstinent, but being confronted with a glass of wine or an option to play your favorite game, you feel unable to resist the desire to perform the act, even if you (later) will regret what you have done. As described above, a series of studies focused on drug effects on dopaminergic neurotransmission suggest that drugs of abuse and potentially also certain games release dramatically more dopamine than natural reinforcers such as food consumption or sexual encounters (figure 8.3) (Di Chiara and Imperato, 1988; Martel and Fantino, 1996; Fuchs et al., 2005).

If drugs of abuse thus release unphysiologically large amounts of dopamine, chronic release of dopamine by drugs of abuse can induce counterregulatory neuroadaptations; for example, a down-regulation of postsynaptic D2 receptors after long-term alcohol use (Rommelspacher et al., 1992; Volkow et al., 1996; Heinz et al., 2004b, 2005c). Likewise, chronic intake of heroin as well as cocaine and other psychostimulants has been associated with a reduction of striatal D2 receptors (Volkow et al., 2001, 2015b; Martinez et al., 2012). At this point, addicted subjects may crave for the drug of abuse not because it still induces an overwhelmingly strong dopaminergic effect, but rather because due to the down-regulation of the dopaminergic system, natural reinforcers have lost their ability to substantially stimulate dopaminergic neurotransmission, while drugs of abuse are still relatively stronger in effect due to their direct pharmacological impact on the dopaminergic system (be it by blocking or even reversing dopamine reuptake or by indirectly stimulating dopamine release). Drugs of abuse may thus be desired because unlike natural reinforcers, they still release some dopamine in the striatum, which is experienced as pleasurable and thus reinforces drug intake (Wise, 1985).

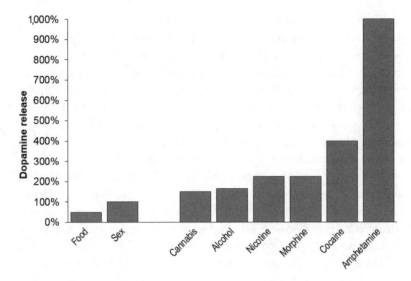

Figure 8.3
Causes of neuroadaption due to drug intake in the dopamine system. Alcohol, like other drugs of abuse, releases dopamine in the striatum. Compared to natural reinforcers such as food consumption, alcohol intake is associated with both a higher amount of dopamine release and a lack of habituation of this neurobiological response; that is, repeated alcohol intake will continue to release relatively high amounts of dopamine in the striatum in animal experiments, thus inducing a counterregulatory down-regulation of striatal dopamine D2 receptors to ensure homeostasis (Rommelspacher et al., 1992; Volkow et al., 1996; Di Chiara and Bassareo, 2007; Heinz et al., 2005c).

An alternative theory has been proposed by Robinson and Berridge (1993), who suggested that chronic intake of drugs of abuse does not down-regulate but rather sensitizes phasic dopamine release, resulting in the well-known phenomenon of increased psychomotor activation after repeated application of amphetamine and other psychostimulants. Such sensitization should also, due to Pavlovian conditioning processes, enhance the effects of drug-associated cues. However, assessing alcohol-dependent subjects during the first weeks of abstinence provided no evidence for sensitization processes but rather consistently showed dopamine D2 receptor down-regulation and reduced dopamine release when stimulated with amphetamine (Volkow et al., 1996; Heinz et al., 2005c; Martinez et al., 2005). Also, imaging studies carried out in recently detoxified alcohol-dependent patients did not show that alcohol-associated cues constantly activate the ventral striatum in addicted patients more than in healthy control subjects, as should be expected in the case where cue-induced dopamine release is sensitized in this brain area (Huys et al., 2016a). Nevertheless, one recent study suggested that after detoxification, there is a delayed up-regulation of dopamine release and dopamine D1 receptors in the striatum of rodents, which manifests about 3 weeks after alcohol intake was stopped; the same group observed an increase in dopamine D1 receptors in postmortem brains of alcohol-dependent patients (Hirth et al., 2016). These observations suggest that sensitization may manifest sometimes after detoxification and could be associated with the speed of recovery of dopaminergic neurotransmission, which may increase even above and beyond levels found in healthy control subjects. Neuroendocrinological challenge studies using apomorphine, a dopamine D1 and D2 receptor agonist known to release growth hormone (GH) via dopamine D2 receptors in the hypothalamus, revealed a reduced sensitivity of dopamine D2 receptors during alcohol intoxication, which recovered within the first days of abstinence in subsequent abstainers, while recovery was delayed in subjects who relapsed during the 6-month follow-up period (Heinz et al., 1996a). Yet again, except for peripheral dopamine markers of doubtful validity, there was no sign of sensitization, and subjects who would later relapse showed a delayed recovery rather than an increased sensitivity of dopamine D2 receptors 1 week and 3 months after detoxification (Dettling et al., 1995; Heinz et al., 1995b). Likewise, there is no clear evidence for sensitized dopaminergic neurotransmission in humans with psychostimulant or heroin dependence, rendering the concept of sensitization plausible yet unconfirmed (Volkow et al., 1997, 2015b; Martinez et al., 2012).

Is there more evidence for sensitization processes in humans when we include further mechanisms such as Pavlovian-to-instrumental transfer (PIT) into our considerations? Indeed, most studies described above simply looked at reactions elicited by the confrontation of addicted subjects with drug-associated cues and did not directly assess the effect of such cues on behavior. If we look at PIT paradigms, the results of such studies may help us to better understand why behavioral tendencies elicited by conditioned cues can be so "compelling." For example, it has been observed that PIT effects, that is, approach-inducing effects of positive (non-drug) Pavlovian cues and withdrawal-augmenting effects of aversive Pavlovian stimuli, are increased in alcohol-dependent patients compared to healthy controls (Garbusow et al., 2014, 2016a). Moreover, the PIT effect was associated with increased ventral striatal activation in alcohol-dependent patients, with the degree of activation being positively correlated with their prospective relapse risk (Garbusow et al., 2016a). These observations may support the hypothesis of a sensitization of dopamine release after detoxification, as suggested by animal experiments and human studies (Robinson and Berridge, 1993; Wassum et al., 2013). Notably, alcohol-associated Pavlovian stimuli increased withdrawal rather than approach tendencies in detoxified alcohol-dependent patients with a good versus poor treatment outcome, and detoxified patients who failed to modulate ventral striatal activation in association with this withdrawal tendency had a higher prospective relapse risk (Garbusow et al., 2016b). This observation indicates that alcohol cues can be experienced as aversive in detoxified patients, which corresponds to verbal reports about disliking alcohol pictures by this patient group. A failure to increase instrumental withdrawal tendencies when confronted with drug cues may reduce caution and increase risk taking in detoxified patients, thus contributing to the prospective relapse risk. An important task may thus be to enable patients to identify cues that predict the availability of drugs of abuse and to control their approach behavior accordingly.

Another hypothesis does not focus on reward dysfunction but rather suggests that cravings are overwhelming because cognitive control is reduced. Cognitive control is often hypothesized to depend on prefrontal cortex functioning, and indeed chronic drug consumption can impair executive behavior control due to neurotoxic effects (e.g., of alcohol on the prefrontal cortex) (Pfefferbaum et al., 2001). However, empirical evidence for such claims is limited: with respect to the development of addiction, in subjects with modest to harmful drinking behavior, a tendency toward harmful drinking was associated with increased functional activation of limbic brain areas associated with reward processing including the ventral striatum, amygdala, and medial prefrontal cortex during decisions favoring a desired alcoholic drink, while there was no impaired activation of brain areas generally associated with executive control including the dorsolateral prefrontal cortex (Stuke et al., 2016). In alcohol dependence, it has been shown that working memory, a key executive function potentially impaired in drug dependence, was rather unimpaired in detoxified alcohol-dependent patients; moreover, patients who later relapsed did not show reduced action of prefrontal and parietal brain areas associated with working memory function (Charlet et al., 2014). Instead, subsequent abstainers but not relapsers appeared to compensate for potential performance challenges by "overactivating" brain areas inside and outside of the neural network that is usually activated when such tasks are performed (Charlet et al., 2014). In contrast, in psychostimulant-addicted subjects, brain activation in the posterior cingulate, right insula, and temporal cortex during a rather simple two-choice prediction task was able to identify subsequent relapses and abstainers (Paulus, Tapert, and Schuckit, 2005), while brain activation patterns in the prefrontal cortex did not strongly predict the relapse risk. These findings suggest that the brain areas identified by Paulus et al. (2005) may well contribute to behavioral control. However, this study does not confirm the hypothesis that executive functions associated with the dorsolateral prefrontal cortex play the key role in controlling drug urges and preventing relapse in detoxified subjects.

Yet another explanation why strong urges and automatic tendencies to consume a drug of abuse can be experienced as irresistible or even "compulsive" is provided by the conditioned manifestation of withdrawal symptoms. Hinson and Siegel (1982) observed that rats show conditioned withdrawal when being placed in a cage in which they regularly received heroin but are not injected with a drug of abuse at this time point. Obviously, counter-regulatory adaptations do not only result from long-term drug intake, but may also manifest as conditioned reactions that are quickly triggered by certain cues or contexts. When trying to assess how frequently such conditioned withdrawal symptoms are observed by detoxified alcohol-dependent patients and whether they subjectively contribute to alcohol craving and relapse, we observed that about one third of our patients described cue- or context-dependent alcohol urges that were associated with typical withdrawal symptoms (Heinz et al., 2003a). One of our patients gave a classical description of such experiences and explained that he got angry when he was rejected at a city office responsible for handing him out his social aid. He stated that in such situations, he would usually have gotten drunk. Being detoxified and having decided to remain abstinent this time, he did not consume alcohol but instead entered a streetcar to go back to our hospital. Suddenly, in the middle of the streetcar and although having been detoxified for more than 10 days, he experienced a reappearance of his withdrawal symptoms: he started to sweat and shiver, trembled, and felt nauseated. These somatic symptoms were accompanied by a strong urge to consume alcohol. Luckily for our patient, the streetcar stopped right in front of the hospital gate, and he was able to run into the ward and ask for help. Clinical experience shows that many patients relapse when withdrawal symptoms appear, be it due to the actual cessation of drug intake or to conditioned processes comparable to the one just described. Unfortunately, conditioned withdrawal has hardly been examined, and adequate scores and test procedures are widely lacking in addiction research.

Altogether, these considerations suggest that different mechanisms and their neurobiological underpinnings are associated with relapse after detoxification including conditioned withdrawal and conditioned drug urges, which have been associated with neuroadaptation to chronic drug intake. But how do such neuroadaptations influence reward-related learning and decision making and thus the behavior toward drug intake?

Drug-Associated Alterations in Learning Mechanisms

Computational accounts of addictive behavior tend to focus on the observation that drugs of abuse release more dopamine than natural reinforcers (Di Chiara and Bassareo, 2007), thus eliciting a stronger error of reward prediction (Redish, 2004). There is also good evidence that conditioned cues that reliably predict reward but are themselves not predicted by preceding stimuli elicit phasic dopamine release to a degree that reflects the error of reward prediction (Schultz et al., 1997), thus attributing incentive salience to drug-associated cues (Robinson and Berridge, 1993; Heinz, 2002a). Drug consumption may thus be preferred because conditioned cues that predict reward elicit strong errors of reward prediction and bias behavior toward drug intake (Redish, 2004).

Robinson and Berridge (1993) further suggest that phasic dopamine release elicited by drugs of abuse is sensitized (i.e., drug intake and drug cues elicit an increasing amount of dopamine). These theories were based on the observation that repeated amphetamine intake sensitizes psychomotor activity, hypothetically due to increased effects on dopaminergic neurotransmission in the striatum. Such sensitization theories may help to explain why reward-related decision making can be strongly biased toward drug seeking and intake. However, sensitization processes with respect to psychomotor activation may be limited to the dorsal striatum and not reflect sensitization processes in more limbic and motivational parts of the basal ganglia. Furthermore, the assumption of sensitized dopamine release in drug addiction should result in increased functional activation elicited by drug-associated cues in the ventral striatum of addicted subjects, which has been found in some but not all imaging studies (Kühn and Gallinat, 2011; Beck et al., 2012; Huys et al., 2016a). In most imaging studies, drug-associated cues indeed activate a series of brain areas including parts of the striatum, the medial prefrontal and anterior cingulate cortex, and the amygdala (Kühn and Gallinat, 2011); however, activation of the ventral striatum is inconsistent and furthermore was surprisingly stronger in alcohol-dependent patients who later remained abstinent compared to those who would relapse during a 3-month follow-up period (Beck et al., 2012). These findings cast doubt on a straightforward account that explains drug seeking and intake by a simple sensitization process.

An alternative theory suggests that drug-induced dopamine release is accompanied by a compensatory down-regulation of dopamine receptors to ensure homeostasis, which impairs dopamine-dependent learning processes and makes it particularly difficult for addicted subjects to change their behavior (Heinz et al., 2016b). Indeed, several imaging studies observed reduced dopamine D2 receptor availability in alcoholism and psychostimulant and heroin dependence (Volkow et al., 1996, 2001; Heinz et al. 2004b, 2005a; Martinez et al., 2012). Drug intake may thus be preferred above and beyond the consumption of natural reinforcers such as food because due to neuroadaptation, all rewards are less reinforcing than before, but drug consumption will still be considerably stronger due to its direct pharmacological effects on dopamine release (Heinz, 2002a). As described above, it was observed that reduced dopamine D2 receptor availability in the ventral striatum was associated with increased processing of drug-associated cues in the medial prefrontal cortex (Heinz et al., 2004b). Furthermore, compensatory down-regulation of dopaminergic neurotransmission in the ventral striatum can impair processing of non-drug-related and rewarding stimuli in this brain area, which was indeed observed in alcohol-dependent patients as well as in pathologic gamblers (Reuter et al., 2005; Wrase et al., 2007). Reduced functional activation of the ventral striatum elicited by rewarding and non-drug-related, reward-predicting stimuli, as observed in these studies, may impair reward-related learning and decision making; however, learning mechanisms were not assessed in these pilot studies.

It took a new wave of computational studies in addiction research to assess interactions between prediction error encoding in the ventral striatum, functional connectivity between the ventral striatum and prefrontal cortex, and behavioral performance in instrumental learning tasks. In the study of Park and coworkers (2010), reduced learning speed was associated with altered functional connectivity between the ventral striatum and frontal cortex during performance of a reversal learning task. Two further studies observed a reduction in goal-directed behavior and a shift toward habitual decision making in drug-addicted patients (Sebold et al., 2014; Voon et al., 2015). In these studies, computational modeling revealed a reduction in goal-directed behavior rather than a direct increase in habitization. This may be surprising given that strong alterations in dopaminergic neurotransmission have been observed in the striatum and not in the prefrontal cortex of addicted subjects, which may more plausibly be associated with impaired goal-directed decision making. However, Deserno et al. (2015b) observed that dopamine synthesis capacity in the ventral striatum modulates error encoding in the prefrontal cortex as well as in the striatum and thus contributes to both goal-directed and habitual decision making, suggesting that dopamine dysfunction in the ventral striatum may indeed impair goal-directed decision making associated with prefrontal information processing.

Clinically, impairment of reward-related learning may be particularly dangerous during early abstinence, when patients have to learn to develop alternative coping strategies instead of resuming drug intake. When unable to acquire such alternative strategies, patients may be particularly drawn toward resuming habitual drug intake, and indeed, impaired functional connectivity between the ventral striatum and the prefrontal cortex during reward-related learning was associated with increased drug craving in alcohol-dependent patients after detoxification (Park et al., 2010).

Habitual decision making may be particularly strong when patients are stressed and confronted with cues indicating drug availability, which can affect instrumental behavior via Pavlovian-to-instrumental transfer. Indeed, it has been observed that subjects who display less model-based control and accordingly mainly rely on model-free decision making also show increased PIT effects (Sebold et al., 2016). There is some first evidence that detoxified alcohol-dependent patients have a higher relapse risk in case they show strong PIT effects when confronted with affectively positive Pavlovian cues (Garbusow et al., 2016a). This observation may indicate that some addicted patients are particularly vulnerable to relapse when confronted with an affectively positive context, hypothetically because drugs of abuse are often consumed in stimulating social contexts, and it has been well documented that not only stressful but also appetitive situations can trigger drug craving and intake (Heinz et al., 2003a).

In this context, individual differences in cue reactivity may contribute to the development and maintenance of addictive behavior: Flagel and colleagues (2008) observed that there are substantial individual differences in animals reacting to the presentation of a conditioned cue that predicts reward: some rodents approached the cue itself, while others directly went to the localization where food will be delivered. The authors also found that animals that approached the cue (so-called sign-trackers) were more sensitive to psychomotor activation elicited by repeated cocaine intake and hence appeared to be more susceptible to cocaine-associated neuroplasticity. Flagel and colleagues (2011) suggested that phasic dopamine release in sign-trackers attributes incentive salience to reward cues and that in these individuals, drug-associated cues can have a profound impact on behavior. These findings may help to explain individual differences in cue-reactivity in addicted subjects.

However, these observations do not help to explain why Beck and coworkers (2012) observed that patients with strong functional activation elicited by alcohol cues in the ventral striatum tended to remain abstinent rather than to relapse during the first 3 months after detoxification, when relapse rates are particularly high in detoxified alcohol-dependent patients (Heinz et al., 1996a). Also, dopamine dysfunction in the ventral striatum after detoxification in alcohol dependence should reduce functional activation elicited by the encoding of reward prediction errors, which was not the case in two independent studies (Park et al., 2010; Deserno et al., 2015b). While reward prediction error encoding was unaltered, bottom-up information processing between the ventral striatum and dorsolateral prefrontal cortex was impaired in detoxified patients compared to healthy controls (Park et al., 2010). Moreover, dopamine synthesis capacity in the ventral striatum was not correlated with reward prediction error encoding in detoxified alcohol-dependant patients (Deserno et al., 2015b), suggesting a loss of dopaminergic fine-tuning of information encoding in this brain region. A study by Garbusow et al. (2016b) may help to explain these findings: Pavlovian-to-instrumental transfer effects elicited by alcohol cues were associated with increased functional activation of the ventral striatum; however, this increased ventral striatal activation was found in patients who would subsequently remain abstinent and not in subsequent relapsers. Notably, alcohol cues decreased rather than increased approach behavior in an unrelated instrumental task, suggesting that alcohol cues are experienced as aversive, in accordance with subjective reports of patients. In detoxified patients, alcohol cue effects on Pavlovian-to-instrumental transfer were thus associated with increased ventral striatal activation as well as increased behavioral inhibition in the instrumental task in subsequent abstainers compared to subsequent relapsers. These observations are in accordance with an incentive salience attribution hypothesis of dopaminergic neurotransmission (Robinson and Berridge, 1993), but they emphasize that salience attribution can occur to both appetitive and aversive stimuli. In fact, in the monetary incentive delay task (Knutson et al., 2001; Heinz and Schlagenhauf, 2010), both cues predicting available reward and stimuli that were associated with the possibility to lose money elicited functional activation of the ventral striatum compared with neutral control cues. Increased ventral striatal activation after the presentation of alcohol cues may thus reflect the aversive impact of such stimuli, and subjects who fail to show inhibitory effects elicited by alcohol cues on instrumental behavior may be particularly in danger to relapse when confronted with available drugs.

Our considerations suggest that computational modeling of reward-related learning as well as Pavlovian-to-instrumental transfer can help to disentangle mechanisms contributing to excessive drug intake as well as relapse in addictive disorders (Redish, 2004; Heinz et al., 2016b). During the development of drug addiction, further factors such as a low level of aversive effects experienced when consuming drugs of abuse (Schuckit and Smith, 1996; Hinckers et al., 2006) can contribute to excessive drug intake, particularly when rewarding effects of drug consumption are relatively high (e.g., due to increased drug effects on endorphinergic neurotransmission) (Heinz et al., 2005b). It has also often been suggested that impulsivity (i.e., the preference of short-term over long-term rewards) may contribute to excessive drug intake and relapse. However, we have pointed out that impulsivity is a vague construct that includes mechanisms such as delay discounting, but also impairments in the ability to monitor motor responses and in emotion regulation (Heinz et al., 2011). Therefore, basic mechanisms implied in impulsive behavior need to be identified and computationally modeled. Learning mechanisms may play a particularly strong role in the development of drug addiction in case there is a genetic disposition toward reduced encoding of the effects of natural reinforcers; however, while there is some evidence in animal models that reduced dopaminergic neurotransmission predisposes to excessive intake of psychostimulants and that high dopamine D2 receptor availability may protect relatives of alcohol-dependent patients from becoming alcohol addicted (Volkow et al., 2006; Thanos et al., 2008), there is so far no confirmation of this hypothesis in longitudinal studies with subjects who later became addicted. Therefore, such longitudinal studies are warranted to assess dopamine neurotransmission and functional correlates of error encoding and reward-related learning, in order to disentangle factors that predispose toward excessive drug intake from neuroadaptive consequences of drug consumption. In spite of these existing knowledge gaps, identification of learning mechanisms can help to individually target therapeutic interventions that have been shown to successfully reduce drug approach and intake (Wiers et al., 2011).

Summary: How We Learn to Become Addicted and What That Tells Us with Respect to Therapy

Altogether, our review of reward learning mechanisms in addiction research suggests that drugs of abuse interfere with neural systems implicated in reward-related learning and decision making and thus bias behavior toward drug intake. Some 25 years ago, animal experiments inspired Robinson and Berridge (1993) to suggest that phasic dopamine release elicited by drugs of abuse is associated with the desire to obtain the drug ("wanting") but not the hedonic pleasure ("liking") associated with its consumption. In accordance with this hypothesis, Schultz and coworkers (1997) reported that the temporally unpredicted presentation of a reward as well as a conditioned, reward-predicting stimulus elicits a phasic increase in dopamine firing, while this was not the case when the reward was fully predicted. These observations have inspired a fundamental shift in addiction theories, which had previously focused on drug reward and pleasure-based reinforcement, toward studying the motivational effects of primary and secondary reinforcers including drugs of abuse (Volkow and Fowler, 2000; Heinz, 2002a).

Inspired by these animal experiments, human studies assessed reward-dependent learning using reversal learning paradigms and observed that encoding of reward prediction errors in the ventral striatum is not impaired, while functional connectivity between the ventral striatum and the frontal cortex is altered, and dopamine synthesis capacity in the ventral striatum failed to interact with functional activation elicited by reward prediction errors in detoxified alcohol-dependent patients (Park et al., 2010; Deserno et al., 2015b). These studies revealed impairments in fine-tuning of the neurobiological correlates of flexible instrumental learning in addiction. A new wave of computational studies further showed that reduced goal-directed decision making is associated with a shift toward habitual responding in drug addiction as well as OCD (Sebold et al., 2014; Voon et al., 2015). However, as already discussed earlier, there are profound differences between OCD and addiction, including different directions of altered fronto-cortical glucose utilization (Baxter et al., 1987, 1992; Volkow et al., 1992, 2015a; Adams et al., 1993). Furthermore, habitual drug intake is initially rewarding, while compulsions are often intended to guard against aversive and threatening obsessions (Heinz, 1999; Schoofs and Heinz, 2013). Also, cues that trigger specific compulsions such as repetitive washing or hoarding have been associated with specific fronto-striatal neurocircuits, which differ from fronto-striatal networks activated by drug and alcohol cues (Kühn and Gallinat, 2011). In addition, habit formation is promoted by mechanisms associated with Pavlovian-to-instrumental transfer, with appetitive Pavlovian stimuli promoting unrelated instrumental approach behavior to a larger degree in detoxified alcohol-dependent patients compared to healthy control subjects (Garbusow et al., 2016a). PIT effects were also evoked by alcohol-associated cues, and PIT-associated functional activation of the ventral striatum predicted future relapse; notably, subsequent abstainers but not relapsers were able to increase behavior inhibition when confronted with subjectively aversive alcohol cues in the unrelated instrumental task (Garbusow et al., 2016b). Failure to inhibit approach to drugs and drug cues thus appears to be an important risk factor for a poor treatment outcome after detoxification, and indeed an alcohol approach bias was observed in detoxified patients and successfully altered with a rather short training program requiring patients to automatically push alcohol cues away (Wiers et al., 2011). In spite of such promising intervention studies, much remains to be explored with respect to interactions between Pavlovian and instrumental learning mechanisms and their modification by environmental factors including stress, their association with neurotransmitter systems, and their malleability by pharmacological or psychotherapeutic interventions (Heinz et al., 2016b). Exploring these interactions and modeling human learning and decision making under different environmental conditions including stress exposure and cue presentation will help to better identify patients in need of special treatment and to individually target interventions aimed to reduce habitual and cue-induced drug intake and to acquire alternative coping skills.

9 Learning Mechanisms in Affective Disorders

If we currently diagnose an affective disorder, available disease categories range from depressive disorders comorbid with anxiety or psychotic symptoms (schizoaffective psychosis) to bipolar disorder (i.e., the periodic manifestation of manic or depressed mood states), and from adjustment disorders with depressed mood to chronic dysthymia and major depression (American Psychiatric Association, 2013; World Health Organization, 2013). When Kraepelin distinguished between severe mental disorders on the basis of typical symptom clusters and clinical course and thus separated supposedly chronic progressive dementia praecox from cyclic manifestations of manic-depressive illness (nowadays schizophrenia and bipolar disorder, respectively), his category of severe affective disorders did not include most of the clinical categories including "dysthymia" or "adjustment disorder with depression" that are currently applied to diagnose specific negative mood states (Kraepelin, 1913). According to Karl Jaspers (1946), both severe affective disorders as well as schizophrenia represent endogenous psychoses; that is, severe diseases whose causes are not yet known (hence labeled as "endogenous") but that nevertheless resemble organic diseases of the brain (exogenous psychoses including Alzheimer's dementia or delirium tremens) with respect to severity and the dynamics of their clinical course. In fact, Jaspers (1946) suggested that there are actually not only two but three categories of so-called endogenous psychoses, the third one being *epilepsia*. This system of categorization helps to understand what Jaspers (1946) and his contemporary colleagues considered as criteria for a severe mental disorder: its manifestation is somehow surprising, not easily explained by social stress factors but instead due to the dynamics of the invisible organic process that triggers symptom manifestation and progression of the disease. Like epileptic seizures, which can manifest "out of the blue," severe depressed or manic mood states can be displayed without adequate environmental stressors explaining the manifestation, let alone the specific phenomena characterizing the disorder.

In contrast, depressed mood states triggered by common psychosocial stress factors were considered to be adjustment disorders or "neuroses"; that is, manifestations of mental suffering due to a combination of a vulnerability that predisposes the subject to a mental disorder (e.g., traumatization or a specific affective disposition or personality structure) and additional events that trigger the manifestation of the acute illness. This clinical typology is reflected in the distinction between "endogenous depression" on the one hand and "variations of human suffering" including neurotic disorders and "reactive" depression on the other (Jaspers, 1946). However, the distinction between "endogenous" and "neurotic" or "reactive" depression was largely abandoned with the establishment of operationalized diagnostic procedures according to the *International Statistical Classification of Diseases and Related Health Problems* (ICD) and the *Diagnostic and Statistical Manual of Mental Disorders* (DSM). Nevertheless, there is some clinical plausibility in this traditional distinction. Consider the following two cases.

A 45-year-old man complained about his 8-week treatment—which consisted in cognitive behavior therapy and the application of antidepressive medication on a ward specialized in treatment of affective disorders—because he did not feel that it had changed his mood "one little bit." The patient had grown up in former East Germany and was a member of the state's ruling Socialist Party and a rather important politician in his district. When the East German state broke down and East and West Germany were reunified in 1990, he did not only lose his political position but all the social recognition and respect that was previously associated with holding a high office in the established social hierarchy of East Germany. At this time, being in his early thirties, he tried to reinvent his career and became an entrepreneur; however, his business failed, he had to claim bankruptcy, and in the economic turmoil, his wife filed for divorce and left their previously common home with her children. At this point, our patient became severely depressed and felt unable to deal with all current demands. He applied for retirement, developed suicidal ideations, was diagnosed with a major depressive disorder, and received inpatient treatment to cope with his clinical depression.

A second patient had always been healthy until the age of 61, when he started what his wife called "breakfast discussions," in which he was increasingly worried about what she felt to be minor problems. She described him to be literally "obsessed" with rather irrelevant decisions and the potential complications and dangers that may result from a wrong choice. He retired, but his anxieties and ruminations increased over the next years. A critical point was reached when he could not decide whether he should attend a high-school class reunion, which was to occur several decades after he had received his high-school diploma. Although he had not seen his classmates for decades and there was no immediate advantage or disadvantage associated with joining the class reunion, he started to continually worry about whether to meet his former classmates or not. His wife was unable to distract him or calm him down. He felt so tortured by this problem that he attempted to commit suicide, first by hanging himself and then by trying to stab himself with a penknife. He stabbed his chest about 40 times, luckily not penetrating much more than skin-deep. When admitted to the hospital, he was hardly able to utter a coherent sentence and was firmly convinced that his attempt to stab himself must have damaged his brain, which he called "his computer." Most of this conversation consisted of the repeatedly uttered phrase "My computer . . . is . . . broken." His movements were slow, and most of the time he was sitting apathetically on a chair, interrupted by short periods of inner agitation. When treated with both supportive psychotherapeutic interventions and a traditional (tricyclic) antidepressive medication, his symptoms did not change for about 14 days. Then suddenly, after a weekend without specific therapeutic contacts, he recovered. On the next Monday, he walked down the central aisle of the ward, greeted his physician with a loud voice, and asked him if he wanted to have a piece of the cake he had baked. He was stable receiving medication and some psychosocial support during the next years; however, major depressive symptoms reoccurred twice when a natural healer suggested stopping antidepressive medication.

The first patient is a classic example of a depressed mood state resulting from psychosocial problems and conflicts that overwhelm the coping capacities of a person. In fact, recovering from his depression would mean that the patient has to face his various problems including having to sell his house to finance his debt and coping with the loss of his wife and children, who refused to be in contact with him. The second patient is an example of what used to be called an "endogenous depression," which manifests without psychosocial stressors that appear severe enough to explain the disorder. In fact, it is hard to understand why being unable to decide whether to attend a high-school class reunion and meeting school friends one has not seen for decades motivated a person to commit suicide or how such common stressors caused the delusional ideation that slightly injuring one's chest could inflict permanent damage to the brain. Karl Jaspers (1946) would have called these two patients "ideal types" ("Idealtypen"), suggesting that they represent prototypical cases of distinguishable disorders. Constructing "ideal types" does not rule out that there are many intermediate phenotypes between such classical examples. However, the symptoms and clinical course of the respective illnesses plausibly suggest that neurobiological causes may play a larger role in the second compared to the first patient. Accordingly, "endogenous depression" due to neurobiological alterations was traditionally clearly distinguished from "neuroses" resulting from external and internal conflicts in everyday life (Jaspers, 1946; Heinz, 2014).

However, the just described distinction between "endogenous" and "neurotic" forms of depression does not work as smoothly as it may appear upon first sight: as his therapists later learned from the second patient's wife, the so-called breakfast discussions of this patient with endogenous features started when his mother fell seriously ill. The patient had a very close relationship with his mother, but when she became ill he felt absolutely unable to visit her and in fact did not see her for several years, during which time she was severely sick and finally died in a senior citizen home. So there was a relevant psychosocial stressor even in this supposedly "endogenous" type of depression, which is expected to be mainly driven by inner biological processes and not by psychologically understandable conflicts or problems. Of course, it is very common in human life that a relative dies, and most people do not experience such severe alterations in mood and drive as encountered by our patient, suggesting that this person was particularly vulnerable to such a loss. However, an interaction between psychosocial stressors and personal vulnerability also characterizes the development of clinical depression in the first clinical case: the patient encountered severely stressful situations when he lost his social status, job, and family after a political system change, yet again not everyone who experiences such rather dramatic alterations in his lifeworld becomes depressed. Observing such interactions between psychosocial stressors and individual vulnerabilities in both "endogenous" and "neurotic" types of depression blurs the line between these traditional categories, and indeed the distinction between endogenous and neurotic depression is no longer upheld with respect to the classification of major depressive disorders in ICD-10 and DSM-5 (American Psychiatric Association, 2013; World Health Organization, 2013). The flipside of lumping together such seemingly heterogeneous types of depression is the lack of a common neurobiological correlate that could be used to guide diagnosis, at least as far as imaging of brain activation patterns is concerned (Müller et al., 2017).

Altogether, these considerations suggest that different types of negative mood states may well have their own flavor and features including individual differences in exposure to environmental stress factors, genetic vulnerabilities, or response to medication (Kirsch et al., 2008; Miura et al., 2014); however, it appears justified to look for basic dimensions associated with key symptoms of major depression. These key symptoms include the depressed mood state, apathy, and the inability to experience joy, as well as further symptoms such as guilt feelings and sleep disturbance (American Psychiatric Association, 2013; World Health Organization, 2013). Such a dimensional approach does not rule out that quite different biological pathways are associated with the manifestation of clinical depression, including thyroid hormone dysfunction (Bauer, Heinz, and Whybrow, 2002), neurobiological effects of social exclusion and poverty in the neighborhood (Heinz et al. 2001; Rapp et al., 2015), alterations in serotonergic neurotransmission (Gryglewksi et al., 2014), or hypothalamic-pituitary-adrenal (HPA) axis alterations observed in major depression or alcohol dependence (Heinz et al., 1996b; Reimold et al., 2010); it just focuses on common final pathways that contribute to key aspects of clinical depression.

However, given the ubiquity of negative mood states, when does it make sense to speak of a mental disorder or even a disease? Indeed, it has been hotly debated whether alterations with respect to classification of grief in DSM-5 versus DSM-IV are helpful or not (Frances, 2013). Answering the question at which point specific negative mood states should be called a clinically relevant depression and whether some forms of depressions should be labeled a "disease" is relevant not only when trying to identify computational correlates of decision making in such mood states but also with respect to the attribution of resources in the mental health care system.

When Clinical Depression Can Be Considered to Be a Disease

According to the criteria to diagnose a mental malady discussed earlier in the text (see chapter 1), mental functions that are generally relevant for human life have to be impaired (the disease criteria), which causes individual harm, be it in the form of suffering or a severe impairment of social participation (the illness and sickness criterion, respectively). So what are the core symptoms of major depression, and do they really represent an impairment of mental functions commonly relevant for human life? According to the World Health Organization (2011), major depression is characterized by depressed mood, apathy, and the inability to experience pleasure (anhedonia).

With respect to the negative mood state, feeling depressed or anxious is a common experience for humans; however, depressed mood is characterized by "being stuck in this feeling" (Holtzheimer and Mayberg, 2011). In the German psychopathologic tradition, this specific aspect of depressed mood was called "rigidity of affect" ("Affektstarre"), suggesting that the afflicted person does not just feel down or blue but is indeed unable to feel otherwise (Bobon, 1983; Heinz, 2014). When we consider whether this key symptom of a major depressive disorder actually represents an impairment of functions relevant for human life and survival, it is evident that consistently being stuck in a depressed mood state (or, in mania, being unable to experience grief even if a beloved person dies) may not directly interfere with human survival; however, it makes it exceedingly difficult to live in a shared world with others ("Mitwelt"; Plessner, 1975).

The second key symptom of a major depressive disorder, anhedonia (i.e., the inability to experience pleasure), has already been discussed in chapter 7, which focused on symptoms occurring in psychotic experiences. In fact, the inability to experience joy occurs in several mental disorders (Heinz et al., 1994) and appears to be closely related to the inability to change your mood even when encountering very positive experiences: if you are "stuck" in the negative mood state, it is indeed implausible that you should be able to experience joy. However, it should be noted that definitions of anhedonia vary, and, comparable to broad concepts such as "impulsivity," different scales to measure anhedonia do not even correlate significantly with each other, suggesting that they do not assess the same phenomenon (Schmidt et al., 2001). This vagueness of the terminus "anhedonia" limits attempts to identify common neurobiological correlates across mental disorders (Heinz et al., 1994). With respect to clinical depression, it appears to be important to distinguish between negative mood states and negative affective states, with mood (in German, "Stimmung") usually referring to longer-lasting emotions, which impact on all experiences by imbuing them with an "atmospheric" emotional background. Mood states thus modify the manifestation of acute affective states such as anger, just like the tuning (in German, "Stimmen") of an instrument modifies the sound of specific notes being played. These considerations suggest that depression is not characterized by the occurrence of negative affective states per se but rather by a persistent reduction in the variability of emotions that can be experienced; in fact, severe depression is often characterized by an inability to experience emotions at all rather than simply feeling sad or down (Bobon, 1983).

Finally, the third key symptom of major depression is apathy, a severe lack of motivation that also occurs in several mental disorders. In severe forms of major depression, patients may wake up in the middle of the night, be unable to fall asleep again, worry extensively about everyday problems, and feel absolutely unable to get up or even just to stop ruminating. Patients with severe apathy do not just move and speak slowly, they also feel that performing even simple acts or making everyday decisions requires a huge effort to overcome what is often described as an "inner inhibition." Apathy may also manifest in the form of profound physical exhaustion and can be accompanied by somatic manifestations such as feeling a huge "pressure" on one's chest that makes it difficult to move and breathe (Jaspers, 1946; Bobon, 1983). Patients suffering from severe apathy may stop eating and drinking, clearly indicating that this key symptom of major depression indeed represents an impairment of a function generally relevant for human life and even directly associated with the survival of the individual.

However, the manifestation of medically relevant symptoms of a disease that are *commonly* relevant for human life does not suffice to diagnose a clinically relevant mental malady—there should be a clear indication that such symptoms actually harm the specific *individual* that encounters them. Therefore, a persistent negative mood state associated with the inability to experience joy or with apathy should only be classified as a clinically relevant mental disorder if it is accompanied by individual suffering or a relevant impairment in social participation. Somebody who will no longer perform activities of daily living such as personal hygiene or food consumption due to severely depressed mood and apathy should hence be diagnosed with a clinically relevant mental disorder. If we follow this line of reasoning, clinically diagnosing a depression depends on an assessment of symptoms and the harm caused by them, independent of how they are caused, be it by thyroid hormone dysfunction or grief (Heinz, 2014). Indeed, the disease classification of the WHO (World Health Organization, 2013) suggests that causation of depression should play no role in clinical diagnosis. The classification system of the American Psychiatric Association, in contrast, used to have a "bereavement exclusion" that did not permit diagnosis of major depression in the first 6 months after the loss of a beloved person except when accompanied by specific symptoms such as suicidality or psychomotor slowing (DSM-IV), whereas in DSM-5, major depression lasting for 2 weeks can be diagnosed even when triggered by grief (American Psychiatric Association, 2013). Indeed, why should a major depression not be diagnosed if it results from witnessing the death of one's own child but can be diagnosed if it results from other psychosocial stress factors such as losing one's job? There is a valid concern that common human experiences such as grief get unduly pathologized when they are mistaken for the manifestation of a major depressive disorder. However, if we follow the above line of reasoning, grief that is accompanied by a persistent negative mood state, anhedonia, and apathy (thus fulfilling the medical criterion to diagnose a disease) but neither causes intense individual suffering (e.g., because the afflicted person feels that such experiences are adequate after the loss of a loved one) nor impairment of activities of daily living (the sickness criterion) should not be considered to be a clinically relevant mental malady. These considerations suggest to take the key symptoms of affective disorders seriously and not to question their validity as medically relevant signs, but it emphasizes that the presence of symptoms fulfilling the disease criterion does not suffice to diagnose a clinically relevant disorder in the absence of personal harm (the illness and sickness criteria). With these cautionary remarks on the diagnostic relevance of medically relevant symptoms in mind, we can now review the neurobiological correlates of depressed mood, anhedonia, and apathy and discuss whether such biological findings affect learning mechanisms that rely on neural processing of aversive and appetitive information.

Common Neurobiological Correlates of Negative Mood States, Anhedonia, and Apathy

If we follow the assumption that positive and negative emotions represent two independent dimensions that classify emotions, depressed mood and the inability to experience pleasure can either result from an increase in negative emotions or a decrease in positive emotions (figure 9.1) (Watson, Clark, and Tellegen, 1988; Russel, Weiss, and Mendelsohn, 1989).

With respect to negative emotions, increased activation of the amygdala and related limbic brain areas has repeatedly been associated with processing of aversive stimuli and can thus contribute to anxiety and depressed mood (Büchel and Dolan, 2000; Nader, Schafe, and Le Doux, 2000). In this context, serotonergic modulation of amygdala activity and further limbic areas appears to play a key role, with a meta-analysis suggesting a 10% reduction in serotonin transporter availability in the midbrain and amygdala of patients with major depression (Gryglewski et al., 2014). Dysregulation of the hypothalamic-pituitary-adrenal axis has repeatedly been observed in many patients with clinical depression (Stetler and Miller, 2011), and increased cortisol levels correlated with low serotonin transporter availability in patients with major depression and alcohol dependence (Heinz et al., 2002; Reimold et al., 2010). Moreover, serotonin transporters are down-regulated after medication with serotonin reuptake inhibitors, so reductions in transporter availability observed in patients who previously received antidepressive medication may reflect both drug effects and disease-related alterations

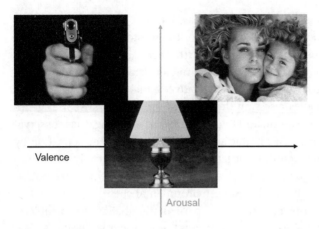

Figure 9.1
Positive and negative affect: mapping affective stimuli according to valence and arousal (cf. figure 5.2 for more detail).

(Benmansour et al., 2002). With respect to clinical correlates, reductions in brain stem serotonin transporter availability correlated with severity of anxiety and depression in alcohol-dependent patients during withdrawal (Heinz et al., 1998c) and with increased anxiety in patients with major depression (Reimold et al., 2008).

Serotonin transporter genotype and availability have been observed to interact with functional activation of the amygdala elicited by aversive stimuli (Hariri et al., 2002; Heinz et al., 2005a; Kobiella et al., 2011). The interaction between serotonin transporter genotype and amygdala activation was partially mediated by effects of serotonin transporter genotype on amygdala volume, pointing to a potential genetic effect during neurodevelopment (Kobiella et al., 2011). In healthy control subjects, the effects of serotonin transporter genotype on amygdala activation appear to be compensated by genotype-driven connectivity between the amygdala and the medial prefrontal cortex, a brain region implicated in emotion regulation: carriers of a specific allele of the promoter of the serotonin transporter (the so-called short allele) displayed less brain stem serotonin transporter availability as well as increased amygdala activation elicited by aversive stimuli, but also increased functional connectivity between the amygdala and medial prefrontal cortex (Heinz et al., 2000a, 2005c; Pezawas et al. 2005). This hypothetically compensatory connectivity between the medial prefrontal cortex and the amygdala was indeed reversed in unmedicated patients with major depression (Friedel et al., 2009). These findings appear to be particularly relevant in the context of a study of Caspi and colleagues (2003), who observed that carriers of the short allele of the serotonin transporter gene, which has previously been associated with decreased protein expression and function (Lesch et al., 1996), have a higher risk to develop depressive symptoms when confronted with severe adverse life events. In light of the neurobiological studies assessing the effect of serotonin transporter genotype on amygdala activation and limbic-prefrontal coupling, these observations suggest that increased limbic and particularly amygdala activation by aversive stimuli contributes to negative mood states, especially when subjects have experienced severe stressful and traumatizing events and emotion regulation fails to cope with aversive environmental cues.

Emotion regulation may indeed be particularly challenged in subjects with a genetic vulnerability that predisposes toward increased limbic activation when confronted with aversive situations as well as toward a potential breakdown of prefrontal-limbic connectivity. Beyond genetic vulnerabilities, emotion regulation can also be challenged in subjects who have experienced severely stressful life events and thus had to learn that their ability to cope with aversive situations is limited. While genetic and environmental risk factors may act independently from each other, the study of Caspi and colleagues (2003) showed that there can be a genotype × environment interaction, with stressful life events having the most profound impact on subjects with a genetic vulnerability toward increased limbic processing of aversive stimuli.

Three arguments can be raised against this explanation of how limbic activation and serotonin dysfunction can contribute to clinical depression. First, these findings are not exclusive to major depression, and serotonin transporter availability was associated with both clinical depression and anxiety in detoxified alcohol-dependent patients and with severity of anxiety in unmedicated patients with major depression (Heinz et al., 1998c; Reimold et al., 2008), indicating that serotonin dysfunction may contribute to a broad range of negative states rather than clinical depression per se. Indeed, serotonergic as well as dopaminergic modulation of amygdala function has repeatedly been implicated in the general processing of aversive stimuli, which can elicit a fear response and contribute to trait anxiety in interaction with further brain areas including the cingulate cortex (Nader et al., 2000; Kienast et al., 2008; Kobiella et al., 2011). Also, serotonin transporter genotype has originally been associated with individual differences in anxiety (Lesch et al., 1996; Sen, Burmeister, and Ghosh, 2004), and clinical depression may only develop when subjects who carry the serotonin transporter genotype associated with increased anxiety and amygdala reactivity are also exposed to severe environmental stressors (Caspi et al., 2003). In this context, it is worth emphasizing that there is considerable overlap not only in the biological correlates of anxiety and depression but also in the clinical phenomena themselves, which is quite plausible regarding the observation that specific feelings of hopelessness and helplessness associated with clinical depression can result from a longer-lasting inability to cope with threatening life situations and the associated anxieties. Indeed, the manifestation of social anxiety disorders often predates clinical depression (Beesdo et al., 2007).

Second, it can be argued that the initial studies measuring the effect of serotonin transporter genotype on amygdala activation observed a rather strong genetic impact in these discovery samples, without which such genotype effects may not have been detected in noisy biological data; however, genotype effects were substantially smaller in subsequent studies as reflected in a meta-analysis, in which serotonin transporter genotype explained about 10% of the variance of amygdala activation (Munafo et al., 2008). Combining the effects of serotonin transporter genotype with further genetic variation that affects monoamine metabolism explained a larger percentage of the variance of limbic activation elicited by aversive stimuli (Smolka et al., 2007), but when information criteria were applied to avoid overfitting of data, the results suggested that in genetic imaging studies, simple models that focus on single polymorphisms may be the most informative; therefore, it is necessary to assess whether an increase in likelihood provided by a complex model (e.g., based on haplotype construction) is large enough to justify the associated increase in model complexity (Puls et al., 2009).

Third, one can critically point out that the interaction between serotonin transporter genotype and environmental stressors with respect to the manifestation of clinical depression has been replicated in some but not in other meta-analyses (Risch et al., 2009; Karg et al., 2011; Sharpley et al., 2014), and other neurotransmitter systems such as dopamine synthesis capacity and metabolism also contribute to functional activation of the amygdala and prefrontal and cingulate cortex (Smolka et al., 2005; Kienast et al., 2008).

Altogether, these considerations suggest that genetic and functional variance in serotonin neurotransmission interacts with limbic processing of aversive stimuli and contributes to negative mood states; however, this contribution seems to be rather unspecific and appears to promote processing of aversive stimuli that can elicit fear, anxiety, as well as depressed mood. With respect to neurobiological models of clinical depression, increased activation of the amygdala and further limbic brain areas has repeatedly been observed in major depression (Mayberg, 1997; Drevets, 2000), suggesting that increased amygdala activation may indeed contribute to negative mood states in clinical depression.

In contrast, Hägele et al. (2016) observed no significant differences in amygdala activation elicited by aversive stimuli in different groups of patients with major depression, schizophrenia, alcohol dependence, attention deficit/hyperactivity disorder (ADHD), or mania compared to each other or to healthy control subjects. This observation suggests that amygdala dysregulation may manifest in specific stressful situations rather than being a trait marker of clinical depression.

With respect to a reduction in positive mood states, it has long been suggested that anhedonia, the inability to experience pleasure, is associated with dopamine dysfunction in major depression and in further severe mental disorders (Heinz et al., 1994). However, these theories were based on the assumption that dopamine mediates the pleasure of obtaining a reward and that anhedonia results from impaired dopamine neurotransmission, be it due to adverse effects of drug consumption or to neuroleptic blockade of dopamine D2 receptors (Wise, 1982, 1985; Heinz et al., 1994). As discussed at length in the previous chapters, our current understanding of dopaminergic neurotransmission suggests that phasic dopamine release encodes positive errors of reward prediction and attributes salience to reward-predicting cues while not contributing to hedonic joy when consuming a reward (Robinson and Berridge, 1993; Heinz, 2002a). This assumption does not rule out that phasic dopamine release, which encodes errors of reward prediction or the availability of reward, also elicits a positive mood state, and indeed Drevets and colleagues (2001) observed that dopamine release elicited by amphetamine in the human ventral striatum was associated with euphoria. These observations suggest that dopamine dysfunction may play an important role in clinical depression, and more specifically that dopamine dysfunction may impair the incentive effects of environmental stimuli and thus contribute to blunted affective responses to reward-predicting cues as well as motivational deficits rather than an inability to enjoy the hedonic pleasure of the consumption of food and other rewards.

In accordance with this hypothesis, Schmidt et al. (2001) observed that reduced dopamine receptor sensitivity was associated with affective flattening rather than anhedonia per se in unmedicated patients with major depression and in further groups of patients with schizophrenia, opiate dependence, and alcohol dependence. Dopamine dysfunction was also implicated in a study of Tremblay and colleagues (2005), who observed that patients with major depressive disorder were hypersensitive to the application of amphetamine with respect to changes in positive mood, which may be due to a compensatory processes induced by a preexisting deficit in dopaminergic neurotransmission in clinical depression. Functional activation of the ventral striatum elicited by reward-predicting cues was reduced in some but not all studies of patients with major depression (Hägele et al., 2015; Misaki et al., 2016). Medication status may play a role in this context, with the application of serotonin reuptake inhibitors normalizing ventral striatal responses toward incentive stimuli (Stoy et al., 2012), in accordance with studies suggesting direct interactions between serotonin 5-HT2A/2C receptor stimulation and striatal dopamine release (Egerton et al., 2008). Confirming alterations in dopamine-rich regions including the brain stem and ventral striatum in major depression, Kumar and coworkers (2008) observed that reward learning signals were blunted in the ventral striatum but increased in the brain stem of patients with major depression, thus pointing to a complex dysregulation of dopaminergic neurotransmission in this clinical disorder.

Beyond dopamine and serotonin, noradrenaline dysfunction has been implicated in major depression. Specifically, depletion of serotonin in patients who previously remitted from clinical depression significantly increased depressed mood, sadness, and hopelessness, while catecholamine depletion impairing noradrenergic and dopaminergic neurotransmission was associated with inactivity, somatic anxiety, and difficulty to concentrate (Homan et al., 2015).

Altogether, these observations suggest that major depression is characterized by an increase in negative mood, which was associated with altered serotonin modulation of processing of threatening cues in limbic brain areas including the amygdala; an impairment in the encoding of reward-predicting stimuli and reward prediction errors, hypothetically due to alterations in dopaminergic neurotransmission that reduce positive mood; and finally apathy and inactivity, which were associated with catecholamine depletion and hence impaired noradrenergic as well as dopaminergic neurotransmission (Heinz et al., 2001; Kumar et al., 2008; Homan et al., 2015).

How Dimensional and Computational Approaches Modify Our Understanding of Major Depressive Disorders

Current models of major depressive disorders suggest that clinical depression is characterized by increased activation of limbic brain areas including the amygdala and reduced activity of dorsal cortical sites including the dorsolateral prefrontal cortex, the dorsal anterior cingulate and posterior cingulate cortex, and the inferior parietal cortex, with a special focus on

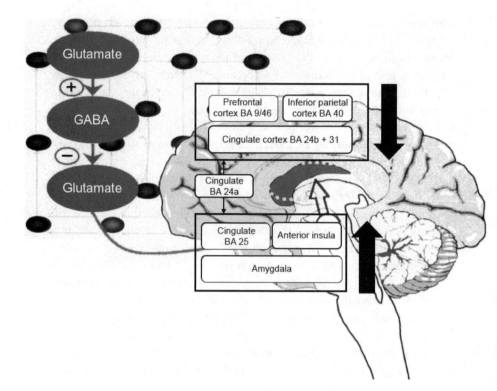

Figure 9.2
Brain areas implicated in major depressive disorder: failure of top-down regulation of limbic and paralimbic brain areas. Clinical depression is supposedly characterized by dysfunction of brain areas associated with cognition and emotion regulation including the prefrontal, parietal, dorsal anterior cingulate, and dorsal cingulate cortex and by overactivity (disinhibition) of limbic and paralimbic brain areas including the subgenual cingulate cortex, anterior insula, and amygdala (Mayberg et al., 1999). These brain areas interact with ascending dopaminergic and serotonergic pathways, which in turn modulate brain activation (Juckel, Mendlin, and Jacobs, 1999). BA, Brodmann area. Source: Modified according to Mayberg (1997).

the rostral anterior cingulate (Brodmann area 24a), which supposedly facilitates the interaction between dorsal and ventral brain areas implicated in major depression (figure 9.2) (Mayberg, 1997).

In this context, some alterations appear to be state dependent while others, particularly in the orbital and medial prefrontal cortex, persist even after symptom remission and are implicated in depression independent of whether mood disturbance follows a unipolar or bipolar course (i.e., independent of whether manic episodes occur or not) (Drevets et al., 1997; Drevets, 2000). Originally, Drevets and colleagues (1997) reported reduced glucose utilization in the subgenual brain area, which would be at odds with the theory that major depression is characterized by increased rather than decreased activity of ventro-limbic and paralimbic brain areas (Mayberg, 1997); however, when applying additional small-volume correction, Drevets and colleagues suggested that glucose utilization in the subgenual cingulate cortex is in fact increased and suggested that unipolar and bipolar affective disorders are associated with dysfunction of an extended brain network that includes the medial prefrontal cortex and further limbic, thalamic, and striatal brain areas (Price and Drevets, 2012). Within this network, the subcallosal cingulate gyrus, which includes Brodmann area (BA) 25 as well as parts of BA 24 and 42, has not only been implicated in unipolar and bipolar affective disorders (Drevets et al., 1997) but also appears to play a key role in mediating the effects of antidepressive therapies (Hamani et al., 2011).

Altogether, visualization of brain areas that are implicated in clinical depression suggests hyperactivation of limbic and paralimbic brain areas including the amygdala and subgenual cingulum and subactivations of parts of the dorsolateral and ventral lateral prefrontal cortex as well as the dorsal anterior cingulate and posterior cingulate cortex (Mayberg, 1997; Drevets, 2000), which interact with monoaminergic neurotransmission (Juckel et al., 1999). These theories resemble assumptions about brain alterations in other major psychiatric disorders. Specifically, it has been suggested that schizophrenia is characterized by dysfunction of the dorsolateral prefrontal cortex, which is supposed to contribute to negative symptoms such as impaired executive functions, and by disinhibition of subcortical dopamine release in the ventral striatum and other limbic brain areas (figure 9.3) (Weinberger, 1987; Heinz and Weinberger 2000; for increased dopamine release in limbic brain areas, see Maia and Frank, 2017).

Such models of brain dysfunction in schizophrenia in turn bear an uncanny resemblance to models of addictive disorders, in which again executive control functions associated with the prefrontal cortex and disinhibited subcortical dopamine release elicited by the consumption of drugs of abuse are supposed to explain reduced executive control and impaired goal-directed behavior on the one hand as well as a disinhibition of drugs of urges and drug-seeking behavior on the other (figure 9.4) (Robinson and Berridge, 1993; Goldstein and Volkow, 2011).

We have already discussed in the past two chapters how computational and dimensional approaches help to diversify these traditional explanatory

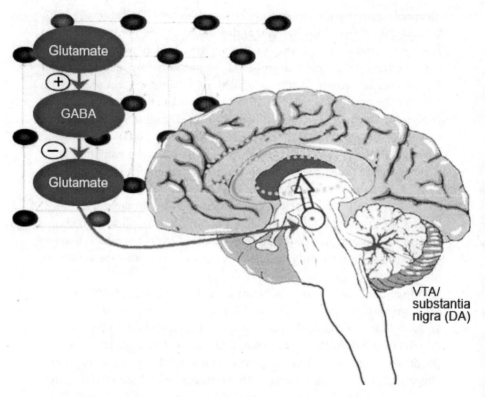

Figure 9.3
Brain areas implicated in schizophrenia: failure of top-down regulation of limbic
and paralimbic brain areas. Schizophrenia is supposedly characterized by failure of
glutamatergic and GABAergic neurotransmission in brain areas associated with cognition and executive behavior control including the prefrontal and anterior cingulate cortex and by overactivity (disinhibition) of limbic and paralimbic brain areas
including the ventral and central striatum (DA, dopamine; VTA, ventral tegmental
area) (Carlsson et al., 1999; Heinz et al., 1999; Kegeles et al., 2000). Source: Modified
according to Heinz et al. (2012b).

models, which appear to be strongly influenced by the suggestion of John
Hughlings Jackson (1884) that mental and neurologic disorders are always
characterized by dysfunction of the highest brain centers and disinhibition
of more primitive brain areas. In contrast to these traditional assumptions,
computational findings have suggested that in schizophrenia, negative
symptoms are associated with impaired signal encoding within the ventral striatum, a core area of the so-called brain reward system (Juckel et al.,

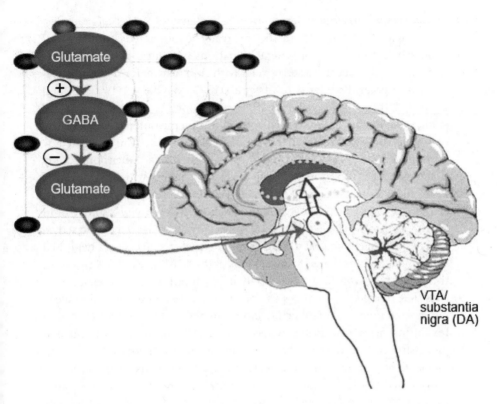

Figure 9.4
Brain areas implicated in addiction: failure of top-down regulation of limbic and paralimbic brain areas including dopaminergic neurotransmission in the ventral striatum. Addiction is supposedly characterized by failure of glutamatergic-GABAergic interactions in brain areas associated with cognition and executive behavior control including the prefrontal and anterior cingulate cortex and by overactivity (disinhibition) of limbic and paralimbic brain areas including the ventral striatum (Berridge and Robinson, 1998; Hyman 2005; Kalivas and Volkow, 2005). Source: Figure modified according to Heinz et al. (2012b).

2006), while delusion formation was associated with altered function of the prefrontal cortex and its top-down influences on sensory information processing (Schlagenhauf et al., 2009; Schmack et al., 2013). In addictive disorders, impaired dopaminergic modulation of ventral striatal error encoding and bottom-up information processing from the ventral striatum to the dorsolateral prefrontal cortex appear to play a key role in learning deficits and drug craving (Park et al., 2010; Deserno et al., 2015a).

With respect to clinical depression, the explanatory models of brain dysfunction we have focused on so far already incorporated a dimensional approach and searched for general alterations in clinical depression independent of whether it was associated with unipolar major depressive disorder or bipolar disorder (Drevets et al., 1997; Mayberg, 1997; Price and Drevets, 2012). Increased limbic brain activation, as implicated in these studies that mainly focused on glucose utilization measured over the course of 1 hour, could be explained by excessive functional activation elicited by aversive stimuli; however, this was not confirmed in a dimensional study of amygdala activation in different groups of patients with major depression, an acute manic episode in bipolar disorder, alcohol dependence, schizophrenia, panic disorder, attention deficit/hyperactivity disorder, or in healthy controls (Hägele et al., 2016). Alternatively, increased limbic activation may be due to neutral stimuli being interpreted as aversive: a depressed subject may thus be biased to always look at the "dark side of life." However, this hypothesis was also not supported by empirical studies: expectancy cues that signaled upcoming positive or aversive pictures elicited reduced rather than increased brain activation in the right lateral orbitofrontal cortex during negative expectancy (Feeser et al., 2013). In contrast, a biased processing of information about upcoming reward or loss was found in mania: here, healthy control subjects displayed increased functional activation elicited by the expectation of loss in the left orbitofrontal cortex, Brodmann areas 11 and 47, while decreased activation was found in the same brain region during the expectation of gain. In patients with mania, activation patterns were reversed, and the same brain region was activated during expectation of increasing gain, while functional activation was decreasing with expectation of increasing loss (Bermpohl et al., 2010). Functional alterations appear to be state dependent and were no longer found after remission of mania (Bermpohl et al., 2010). Altogether, these observations indicate that clinical depression is characterized by impaired processing of expected value of gains and losses in the orbitofrontal cortex, which may be associated with specific alterations in resting-state connectivity observed in bipolar disorder versus unipolar depression (He et al., 2016; for alterations in resting-state connectivity in schizophrenia, see Pankow et al., 2015).

Beyond functional activation elicited by cues that indicate upcoming reward or loss of different magnitudes, studies focusing on learning mechanisms in clinical depression observed reduced encoding of reward prediction errors in the ventral striatum (Kumar et al., 2008); however, currently there is still no unifying computational theory of impaired reward learning in clinical depression (Chen et al., 2015). Some consistent evidence suggests that Pavlovian conditioned cues fail to elicit effects on instrumental behavior as assessed by Pavlovian-to-instrumental transfer tasks: Huys et al. (2016b) observed that in healthy controls, appetitive Pavlovian conditioned stimuli promoted increased instrumental activity to gain a reward and aversive conditioned stimuli facilitated active withdrawal in the unrelated instrumental task, while no such Pavlovian-to-instrumental transfer effects were observed in depressed patients (Huys et al., 2016b). Furthermore, conditioned stimuli associated with emotionally salient versus neutral pictures failed to activate the dorsal medial prefrontal cortex in patients with major depression compared to healthy control subjects (Bermpohl et al., 2009). Finally, conditioned stimuli signaling whether an emotionally salient versus neutral picture would follow activated the dorsomedial prefrontal cortex in healthy control subjects, and again this modulatory effect of conditioned cues on brain activation was not found in depressed subjects (Bermpohl et al., 2009). Together with the above-mentioned observation that cues eliciting expectancy of upcoming aversive stimuli induced a deactivation rather than a hyperactivation of the right lateral orbitofrontal cortex in unmedicated patients with major depression (Feeser et al., 2013), these observations suggest that Pavlovian conditioned stimuli may fail to elicit action-specific effects in clinical depression. So far, we have discussed Pavlovian-to-instrumental transfer (PIT) effects mainly within the context of addiction, where affectively positive Pavlovian conditioned cues stimulate instrumental behavior that can be detrimental when it promotes drug taking and contributes to relapse. However, this focus on PIT effects may altogether be too negative, and there can be many situations in which Pavlovian conditioned cues are indeed helpful in modulating or biasing instrumental behavior. Indeed, PIT effects as well as habits can help to simplify decision trees and may be specifically helpful only in stressful situations, when decisions have to be fast and profit from an overall "atmospheric" evaluation of the dangerousness or safety of the current situation. Beyond stressful contexts, PIT effects may also be useful in inhibiting instrumental activities when there are signs of upcoming danger. Likewise, PIT effects can help to promote instrumental activities when Pavlovian conditioned cues signal that overall conditions are positive. Apathy and hopelessness in clinical depression may thus not simply arise from an increase in negative mood states and a decrease in the encoding of expected values and upcoming reward but also be influenced by a certain failure of specific or general contextual cues to modify behavior accordingly.

These considerations bring us back to the idea that at the core of depressive disorders is not sadness or some other kind of negative mood state per se, but rather the inability to disengage from this negative mood state (i.e., due to the rigidity and inflexibility of this affective state and its neurobiological correlates). This rigid negative mood state is subjectively characterized by an inability to tune in with the "atmospheric" emotional "vibrations" of the environment, and a profound lack of energy associated with the feeling that even simple actions can only be carried out by overcoming some kind of inhibition that weighs down heavily on the afflicted subject. If we follow the suggestion of Martin Heidegger (2006) that our being-in-the-world is always deeply imbued by our current mood state, then feeling as if rigidly frozen in mania or depression can profoundly interfere with human interactions. A failure of contextual cues to modulate approach and withdrawal tendencies may well contribute to these profound emotional and motivational alterations. Learning mechanisms including Pavlovian and operant conditioning can thus help to identify core mechanisms in affective disorders. Such considerations do not contradict the importance of complex alterations in goal-directed decision making embedded in subjective perception of space and time in major depression (Adams, Huys, and Roiser, 2016). Instead, they suggest that beyond alterations in top-down processes associated with cognition and emotion regulation, alterations in basic learning mechanisms and Pavlovian-to-instrumental transfer as well as impairments in functional activation elicited by conditioned cues can have a profound impact on motivation and mood. These considerations emphasize the importance of bottom-up processes such as dopaminergic neurotransmission in the ventral striatum and its impact on the encoding of reward-predicting stimuli as well as Pavlovian-to-instrumental transfer effects (Schultz et al., 1997; Wassum et al., 2013). In light of the complex networks implicated in the manifestation of depressed mood (Mayberg, 1997; Price and Drevets, 2012), a focus on monoaminergic bottom-up influences on Pavlovian and instrumental learning mechanisms and their respective contribution to neural processing of appetitive and aversive stimuli can help to identify similarities and differences between different mood disorders and help to individually target behavioral as well as pharmacological interventions.

10 How a Focus on Learning Mechanisms Can Change Our Understanding of Mental Disorders

Our review of key mechanisms contributing to the development and manifestation of major mental disorders provided substantial evidence that alterations in learning mechanisms help to explain key symptoms of schizophrenia and related psychotic experiences, unipolar and bipolar affective disorders, addictions, and obsessive-compulsive disorders. Specifically, functional effects of Pavlovian conditioned, drug-related cues and Pavlovian-to-instrumental transfer are increased in addicted patients and contribute to drug seeking and drug consumption, while reduced Pavlovian-to-instrumental transfer effects in major depression may contribute to motivational deficits in this disorder (Garbusow et al., 2014; Huys et al., 2016b). Dopamine-dependent attribution of salience to otherwise irrelevant stimuli contributes to delusion formation in schizophrenia and related psychotic disorder (Heinz, 2002a; Kapur, 2003; Schlagenhauf et al., 2014), while noisy information processing in psychotic episodes and counterregulatory neuroadaptation to the effects of drugs of abuse help to explain impaired encoding of reward prediction errors and reward-dependent learning in schizophrenia and addiction (Heinz and Schlagenhauf, 2010; Deserno et al., 2015b; Heinz et al., 2016b). Finally, alterations in serotonergic modulation of limbic processing of aversive stimuli contribute to anxiety and related negative mood states in major depression and in obsessive-compulsive disorder (Heinz, 1999; Reimold et al., 2008; Kobiella et al., 2011).

While these findings support the assumption that learning mechanisms contribute to the phenomenology of major psychiatric disorders, there is substantial interindividual variance among patients and control subjects, and the emerging picture is less clear than may have been expected. Particularly, simple accounts of learning mechanisms may have assumed that negative mood states are characterized by increased Pavlovian-to-instrumental transfer effects elicited by aversive but not appetitive stimuli across mental disorders, which was not the case (Huys et al., 2016a). Also, learning from punishment was not simply increased in disorders characterized by negative mood states; rather, several psychiatric disorders were characterized by impaired reward-related learning as observed in reversal learning tasks that also assess flexible behavior control, including major depression (Kumar et al., 2008), alcohol dependence (Park et al., 2010; Deserno et al., 2015b), and schizophrenia (Schlagenhauf et al., 2013). These findings support a dimensional approach that tries to identify basic neurobiological mechanisms contributing to alterations in subjective experience and objective decision making of patients with major psychiatric disorder. However, our review of the current state of knowledge shows that learning mechanisms and their alterations may be as complex as our current psychiatric classifications appear to be.

Clinical diagnosis of a mental disorder depends on the identification of key symptoms, which indicate an impairment of mental functions generally relevant for human life, and the individual harm associated with their manifestation (Heinz, 2014). Definition and diagnosis of such clinically relevant key symptoms of a disease result from a process of complexity reduction and abstraction, which is inherent in all scientific endeavors (Kuhn, 1962; Fleck, 1979). With respect to clinical practice, complexity reduction and abstraction is aimed at practical values including prognosis and recommendations for suitable therapies. Accordingly, the basic structure of our current diagnostic systems is largely based on observations of symptom clusters and their longitudinal course (Bleuler, 1911; Kraepelin, 1913; American Psychiatric Association, 2013; World Health Organization, 2011). However, the widespread application of pharmaceutical drugs such as antidepressive and antipsychotic medication has helped to question existing categorical systems in psychiatry: antidepressive medication is generally applied for all negative mood states, independent of whether they manifest in major depression, anxiety disorders, drug addiction, or obsessive-compulsive disorders, and antipsychotic medication is not restricted to schizophrenia but is also used in the treatment of delirium and other so-called organic brain disorders. The respective neurotransmitter system targeted by these medications, the serotonin and dopamine systems, have accordingly long been in the focus of dimensional approaches in biological psychiatry (van Praag, 1967; van Praag et al., 1990; Heinz et al., 1994). Dimensional approaches have accordingly been pursued long before the National Institute of Mental Health officially supported trans-diagnostic approaches in neurobiological research of mental disorders (Insel et al., 2010). Again, while there is convincing evidence that dopaminergic and serotonergic neurotransmission play an important role in the encoding of appetitive and aversive stimuli and their respective effects on learning and decision making (Montague, Hyman, and Cohen, 2004; Di Chiara and Bassareo, 2007; Cools, Nakamura, and Daw, 2011), the emerging picture is complex and suggests that dopamine is not only involved in the encoding of appetitive stimuli in the ventral striatum but also modulates amygdala activation elicited by aversive cues (Kienast et al., 2008). Also, serotonergic neurotransmission not only interacts with amygdala activation elicited by aversive stimuli but also stimulates endorphin release in the ventral striatum, thus contributing to positive reinforcement (Zangen, Nakash, and Yadid, 1999; Heinz et al., 2001; Kobiella et al., 2011).

Do Computational Models Help to Clarify the Picture?

Without doubt the work of Schultz and colleagues (1997), which identified dopamine's role in error encoding and provided a computational framework to explain these effects, has stimulated a wide and productive field of research on learning mechanisms in mental disorders. Different ways of decision making can be individually modeled, and comparison of model fit between different computational accounts of behavior can help to identify the computations that a certain individual will most likely have performed when coping with a certain task (Stephan et al., 2016). Within a Bayesian framework, such computational approaches aim at providing a generative model that describes the computational steps taken by the brain when confronted with certain tasks in the environment (Friston et al., 2014; Stephan et al., 2016). However, computations in the brain may again be ultimately more complex, with only some key processes captured by the best-fitting computational model. Nevertheless, such approaches promise to directly link neurobiological processes with a computational understanding of decision making. Today, the most impressive evidence for this approach is still the identification of reward prediction errors that are encoded in the ventral striatum and modulated by dopaminergic neurotransmission (O'Doherty et al., 2003; Schlagenhauf et al., 2013), findings that directly reflect computational roles attributed to dopamine discharge in an animal experiment (Schultz et al., 1997). Again, dopamine's role in information processing appears to be ultimately more complex, with some features of dopamine discharge encoding uncertainty (Fiorillo, Tobler, and Schultz, 2003) and others contributing to motivation (Robinson and Berridge, 1993) as well as positive mood states (Drevets et al., 2001). Computational approaches offer a searchlight in the dark that helps to identify certain key mechanisms implicated in reward learning and salience attribution in mental disorders. If we stay modest and do not claim that what we find is the whole picture, such approaches can help to better understand basic mechanisms of mental disorders and—by identifying individual differences—to better target our psychotherapeutic as well as pharmaceutical interventions.

Do Genetic Studies Help to Specify the Role of Learning Mechanisms in Mental Disorders?

Most major mental disorders are substantially heritable, with twin studies suggesting that about one third to two thirds of the variance of disease manifestation can be explained by genetic factors (Johnson, van den Bree, and Pickens, 1996; Bienvenu, Davydow, and Kendler, 2011). However, we should keep in mind that such assessments of heritability rates assume additive genetic effects, which may not be the case when genetic variance affects, for example, both neurotransmitter synthesis capacity as well as receptor sensitivity; in such cases, increased similarity between monozygotic twins may result from superadditive genetic effects, and twin studies may tend to overestimate heritability rates. Moreover, traditional twin studies cannot distinguish between genetic and epigenetic effects, with first evidence pointing to a role of epigenetic alterations associated with environmental stressors in the development of mental disorders (Swartz, Hariri, and Williamson, 2017). In this landmark study, low socioeconomic status during adolescence was associated with increased methylation of the promoter of the serotonin transporter gene, which in turn was correlated with greater amygdala activation elicited by threat-related stimuli (Swartz et al., 2017). The authors further showed that increased amygdala activation moderated the association between a family history that was positive for clinical depression and the later manifestation of depressive symptoms in the study population. Such pilot studies require replication; however, they suggest that the observed association between socioeconomic adversity and negative mood states (Rapp et al., 2015) may at least partially be due to environmental effects on epigenetic mechanisms, which modulate the functions of brain areas that process aversive stimuli and contribute to learning from negative outcomes. The study of Swartz and coworkers (2017) complements previous findings regarding the effects of genetic variance in the promoter of the serotonin transporter on limbic activation and functional connectivity elicited by aversive stimuli and on the risk to develop clinical depression (Caspi et al., 2003; Heinz et al., 2005a; Kobiella et al., 2011).

Genotype effects have also been observed to interact with ventral striatal sensitivity for reward value as well as prefrontal activity evoked by anticipation of reward (Yacubian et al., 2007), with functional activation associated with working memory performance (Egan et al., 2001) and with prefrontal and limbic activation elicited by affective stimuli (Smolka et al., 2005, 2007). However, assessing more and more complex gene-gene interactions including haplotype construction carries the risk of overfitting the data, and model comparison is required to assess whether the increase in likelihood that is provided by more complex genotype models indeed justifies the increase in model complexity (Puls et al., 2009).

Given the polygenetic nature of our psychiatric disorders, with single mutations contributing only a tiny fraction to the explained variance and considerable overlap between different mental disorders including schizophrenia and bipolar disorder (International Schizophrenia Consortium et al., 2009; Schizophrenia Working Group of the Psychiatric Genomics Consortium, 2014), it is not likely that the identification of specific polymorphisms and their functional impact will help to identify more than a rather limited contribution to neurobiological alterations in key mental disorders. As a case in point, genetic variance in the promoter region of the serotonin transporter appears to modify limbic activation elicited by aversive cues; however, a meta-analysis revealed that the overall effect is considerably smaller than originally described in the discovery samples (Munafo, Brown, and Hariri, 2008), and epigenetic effects associated with environmental factors including socioeconomic status may further complicate the picture (Swartz et al., 2017).

Altogether, these considerations suggest that when trying to understand mental disorders, we should distinguish between different levels of evidence and conceptualization, which are all characterized by high complexity. On a basic level, the effects of genetic and epigenetic variance are assessed with respect to protein expression including neurotransmitter receptors and transporters; the next conceptualization tries to capture overall effects of genetic as well as environmental factors on neurotransmitter systems that interact with key learning mechanisms and thus modulate information processing, decision making, as well as memory formation in neural networks; finally, computational models of decision making can help to identify the neurobiological correlates of key computational steps including the encoding of reward prediction errors in decision making and thus help to reveal individual differences in decision making.

But even when constructing such computational models of behavior, we should never forget that all human beings react to what they experience, and that individual behavior reflects not only alterations in learning mechanisms but also compensatory and creative efforts of individuals to cope with altered and often surprising or threatening experiences. For example, schizophrenia patients are relatively better than control subjects in identifying behavioral relevance of previously neutral cues (Lubow and Gewirtz, 1995), showing that even potentially dopamine-related alterations in salience attribution can have a "creative side" by facilitating the notion of a change in relevance of patterns that most people tend to ignore. A focus on learning mechanisms thus emphasizes the malleability and creativity of human behavior.

Beyond Mental Maladies: Disorders Associated with Common Life Events and Individual Personality

Traditionally, so-called neurotic disorders associated with aversive or even traumatizing life events as well as personality disorders were not regarded as "true diseases" (Jaspers et al., 1946). Currently, both the American Psychiatric Association and the World Health Organization avoid discussions of the disease status of such phenomena and instead speak of "disorders" (American Psychiatric Association, 2013; World Health Organization, 2011). If we follow the suggestion to only regard a state of suffering as a clinically relevant mental malady if there are symptoms of a disease accompanied by either individual suffering or a severe impairment of social participation, everything depends on whether or not we regard key symptoms of any given disorder as impairments of functions generally relevant for human life. We have previously argued (see chapter 1) that the diversity of human behavior is too complex for simple normative accounts and that the discussion of whether certain symptoms can be regarded as indicators of a medical disease depends on whether the impaired functions are necessary for individual survival or at least for living with others in a common world ("Mitwelt" in the sense of Plessner, 1975). We have suggested that manifestation of the symptom of "disorientation in space or time" can directly be life threatening, while the "inability to experience pleasure" in severe depression or the "inability to mourn when confronted with the loss of a beloved person" in mania is not directly relevant for individual survival but can severely impair living with others. With respect to the multitude of disorders listed in current nosological classifications, the key symptoms of each disorder have to be assessed according to whether they indicate an impairment of mental functions that are indeed generally relevant for living with others. Suffering from so-called borderline personality disorder with frequent dissociative states after severe traumatization in childhood may certainly fulfill this disease criterion, while being unable to present lectures in front of a large audience in some forms of social phobia may not be regarded as indicating an impairment of a functional ability that is generally relevant for human life. In this context, we need to emphasize that there is no absolute criterion that distinguishes functions that are commonly relevant for human life from those that are not: for a teacher or a university lecturer, it may be essential to be able to speak in front of large audiences, while this may not be the case for a metalworker or programmer. Indeed, the decision of whether a certain mental disorder fulfills disease criteria is based on plausibility, not on objective measures (Heinz, 2014).

Nosological classifications are based on complex public communication processes that are aimed at ensuring consensus, and we suggest that whether or not a certain impairment can actually count as a symptom of a disease should be openly discussed, not only by professionals but also by patients and their relatives as well as an interested public. We feel that it is important to not label all kinds of suffering associated with mental functioning a disease. On the one hand, labeling all kinds of common problems a disease can increase the pressure on individuals to perform: if every suffering that frequently occurs in everyday life is labeled a disease, employers can expect their employees to immediately get adequate medical treatment instead of just being unable for some time to perform at a certain level of expectation. Moreover, subjects suffering from such common impairments may feel forced to take medication or to otherwise increase their output rather than questioning whether the job pressure or social conditions are altogether inadequate. Social problems could thus be individually pathologized and treated instead of evoking solidaric political action.

Moreover, in many countries the disease status of any mental impairment is a necessary prerequisite for considering compulsory treatment. For example in Germany, a person can only be treated against her current will if she suffers from a clinically relevant mental malady that endangers herself or others and if she also lacks insight into these dangers associated with her current behavior (Bundesverfassungsgericht, 2011). If every impairment of one of the multifold mental functions were to suffice to constitute a clinically relevant mental malady, then compulsory treatment could be extended to a substantial part of the population that shows behavior patterns endangering themselves. However, this appears to be wrong: human beings engage in all kinds of risky behavior, from parachuting to mountain climbing or speed driving, and while such preferences may well be associated with certain personality traits and even a certain lack of insight in the danger that these activities pose to the individual, we strongly feel that any kind of compulsory psychiatric treatment should only be considered if self-endangering behaviors can be attributed to a medically relevant malady; that is, a state characterized by the impairment of one or several mental functions that are generally relevant for human life and survival and that individually cause harm to the afflicted person, be it because of suffering or a severe limitation of social participation (Heinz, 2014). Accordingly, a patient with mania who engages in all kinds of risky behavior fulfills the disease criterion only because the inability to feel grief is commonly regarded as an impairment of a basic mental function relevant for living with others. However, diagnosing a disease does not suffice to consider compulsory treatment: this should only be done in case such manic syndromes are actually harmful for the individual and are associated with a lack of insight into the dangers associated with the manically disinhibited behavior. In neurology, compulsory retreatment is limited to severe brain-organic disorders characterized by key neurologic dysfunctions relevant for humans (e.g., dysfunctions of sensory or motor performance or coordination), which also impair insight and endanger the health of the individual. In accordance with such procedures in somatic medicine, compulsory treatment in psychiatry should, as we feel, only be considered in patients who (1) display a limited number of diseases, in which (2) basic mental functions commonly relevant for survival are impaired and cause (3) individual suffering or interfere with social participation by impairing common activities of daily living (e.g., in Alzheimer's disease, impaired activities of daily living include the inability to keep up personal hygiene or to consume food), and who (4) also lack insight into the dangers posed by their current behavior, which (5) is indeed endangering the health of the afflicted person or of others. Limiting the number of disorders that actually count as diseases is thus a way to limit compulsory treatment in psychiatry, in accordance with the United Nations Convention on the Rights of Persons with Disabilities (United Nations, 2006; Méndez, 2014).

In this book, we do not discuss such states beyond the mental disorders that were classically understood as diseases. This does not rule out that some or even many of the states of suffering that are listed in current nosological classification systems fulfill disease criteria, it just calls for an open debate on whether or not they should be understood as diseases. This debate has to be public and should necessarily involve patients and relatives, who are the ones most directly affected by such decisions. Identifying basic learning mechanisms that contribute to mental maladies and considering their general as well as individual impact on human life can thus help not only to better understand mental disorders and to individually adjust treatment but also to limit compulsory treatment and to promote person-centered therapeutic approaches that focus on the malleability and creativity of human behavior.

References

Abercrombie ED, Keefe KA, DiFrischia DS, Zigmond MJ. 1989. Differential effect of stress on in vivo dopamine release in striatum, nucleus accumbens, and medial frontal cortex. *J Neurochem 52*: 1655–1658.

Abi-Dargham A, Rodenhiser J, Printz D, Zea-Ponce Y, Gil R, Kegeles LS, Weiss R, et al. 2000. Increased baseline occupancy of D2 receptors by dopamine in schizophrenia. *Proc Natl Acad Sci USA 97*: 8104–8109.

Abi-Dargham A, Mawlawi O, Lombardo I, Gil R, Martinez D, Huang Y, Hwang DR, et al. 2002. Prefrontal dopamine D1 receptors and working memory in schizophrenia. *J Neurosci 22*: 3708–3719.

Adams RA, Huys QJM, Roiser P. 2016. Computational psychiatry: Towards a mathematically informed understanding of mental illness. *J Neurol Neurosurg Psychiatry 87*: 53–63.

Adams KM, Gilman S, Koeppe RA, Kluin KJ, Brunberg JA, Dede D, Berent S, Kroll PD. 1993. Neuropsychological deficits are correlated with frontal hypometabolism in positron emission tomography studies of older alcoholic patients. *Alcohol Clin Exp Res 17*: 205–210.

Adams RA, Stephan K, Brown H, Frith C, Friston K. 2013. The computational anatomy of psychosis. *Front Psychiatry 4*: 47.

Akam T, Costa R, Dayan P. 2015. Simple plans or sophisticated habits? State, transition and learning interactions in the Two-Step Task. *PLOS Comput Biol 11*: e1004648.

Alexander GE, DeLong MR, Strick PL. 1986. Parallel organization of functionally segregated circuits linking basal ganglia and cortex. *Annu Rev Neurosci 9*: 357–381.

American Association of Physical Anthropologists (AAPA). 1996. Statements on biological aspects of race. *Am J Phys Anthropol 101*: 569–570.

American Psychiatric Association. 2000. *Diagnostic and Statistical Manual of Mental Disorders, DSM-IV. Text Revision (DSM-IV-TR)*. American Psychiatric Publishing, Washington, DC.

American Psychiatric Association. 2013. *Diagnostic and Statistical Manual of Mental Disorders, Fifth Edition (DSM-5)*. American Psychiatric Publishing, Washington, DC.

Anderson GM, Barr CS, Lindell S, Durham AC, Shifrovich I, Higley JD. 2005. Time course of the effects of the serotonin-selective reuptake inhibitor sertraline on central and peripheral serotonin neurochemistry in the rhesus monkey. *Psychopharmacology 178*: 339–346.

Andreasen NC. 1982. Negative symptoms in schizophrenia. Definition and reliability. *Arch Gen Psychiatry 39*: 784–788.

Andreasen NC. 1990. Positive and negative symptoms: Historical and conceptual aspects. In: Andreasen NC (Ed.), *Positive and Negative Symptoms and Syndromes*, pp. 1–42. Modern Problems of Pharmacopsychiatry, Vol. *24*. Karger, Basel.

Asensio S, Romero MJ, Romero FJ, Wong C, Alia-Klein N, Tomasi D, Wang GJ, et al. 2010. Striatal dopamine D2 receptor availability predicts the thalamic and medial prefrontal responses to reward in cocaine abusers three years later. *Synapse 64*: 397–402.

Atzil A, Touroutoglou A, Rudy T, Salcedo S, Feldman R, Hooker JM, Dickerson BC, Catana C, Feldman Berett L. 2017. Dopamine in the medial amygdala network mediates human bonding. *Proc Natl Acad Sci U S A 9*: 2361–2366.

Bandelow B, Baldwin D, Abelli M, Altamura C, Dell'Osso B, Domschke K, Fineberg NA, et al. 2016. Biological markers of anxiety disorders, OCD and PTSD—a consensus statement. Part I: Neuroimaging and genetics. *Biol Psychiatry 17*: 321–365.

Barch DM, Bustillo J, Gaebel W, Gur H, Heckers S, Malaspina D, Owen MJ, et al. 2013. Logic and justification for dimensional assessment of symptoms and related clinical phenomena in psychosis: Relevance to DSM-5. *Schizophr Res 150*: 15–20.

Bart G, Kreek MJ, Ott J, LaForge KS, Proudniko D, Pollak K, Heilig M. 2005. Increasing attributable risk related to a functional mu-opioid receptor gene polymorphism in association with alcohol dependence in central Sweden. *Neuropsychopharmacology 30*: 417–422.

Bauer M, Heinz A, Whybrow PC. 2002. Thyroid hormones, serotonin and mood: Of synergy and significance in the adult brain. *Mol Psychiatry 7*(2): 140–156.

Baumgarten HG, Grozdanovic Z. 1985. Psychopharmacology of central serotonergic systems. *Pharmacopsychiatry 18*: 180–187.

Baxter LR, Jr., Phleps ME, Mazziotta JC, Guze BH, Schwartz JM, Selin CE. 1987. Local cerebral glucose metabolic rates in obsessive-compulsive disorder. A comparison with rates in unipolar depression and in normal controls. *Arch Gen Psychiatry 44*: 211–218.

Baxter LR, Jr., Schwartz JM, Bergman KS, Szuba MP, Guze BH, Mazziotta JC, Alazraki A, et al. 1992. Caudate glucose metabolic rate changes with both drug and behavior therapy for obsessive-compulsive disorder. *Arch Gen Psychiatry 49*: 681–689.

Bayes T. 1763. An essay towards solving a problem in the doctrine of chances. Communicated by Mr. Price, in a letter to John Canton. *Philos Trans 53*: 370–418.

Bechara A. 2003. Risky business: Emotion, decision-making, and addiction. *J Gambl Stud 19*: 23–51.

Beck A, Wüstenberg T, Genauck A, Wrase J, Schlagenhauf F, Smolka MN, Mann K, Heinz A. 2012. Effect of brain structure, brain function, and brain connectivity on relapse in alcohol-dependent patients. *Arch Gen Psychiatry 69*: 842–852.

Beesdo K, Bittner A, Pine DS, Stein MB, Höfler M, Lieb R, Wittchen U. 2007. Incidence of social anxiety disorder and the consistent risk for secondary depression in the first three decades of life. *Arch Gen Psychiatry 64*: 903–912.

Bel N, Artigas F. 1993. Chronic treatment with fluvoxamine increases extracellular serotonin in frontal cortex but not in raphe nuclei. *Synapse 15*: 243–245.

Belin D, Mar A, Dalley J, Robbins T, Everitt B. 2008. High impulsivity predicts the switch to compulsive cocaine-taking. *Science 320*: 1352–1355.

Belin-Rauscent A, Daniel ML, Puaud M, Jupp B, Sawiak S, Howett D, McKenzie C, Caprioli D, Besson M, Robbins TW, Everitt BJ, Dalley JW, Belin D. 2016. From impulses to maladaptive actions: the insula is a neurobiological gate for the development of compulsive behavior. *Mol Psychiatry 21*: 491–499.

Benmansour S, Owens WA, Cecchi M, Morilak DA, Frazer A. 2002. Serotonin clearance in vivo is altered to a greater extent by antidepressant-induced downregulation of the serotonin transporter than by acute blockade of this transporter. *J Neurosci 22*: 6766–6772.

Bermpohl F, Walter M, Sajonz B, Lücke C, Hägele C, Sterzer P, Adli M, et al. 2009. Attentional modulation of emotional stimulus processing in patients with major depression—alterations in prefrontal cortical regions. *Neurosci Lett 463*: 108–113.

Bermpohl F, Kahnt T, Dalanay U, Hägele C, Sajonz B, Wegner T, Stoy M, et al. 2010. Altered representation of expected value in the orbitofrontal cortex in mania. *Hum Brain Mapp 31*: 958–969.

Berridge KC, Robinson TE. 1998. What is the role of dopamine in reward: Hedonic impact, reward learning, or incentive salience? *Brain Res Brain Res Rev 28*: 309–369.

Bienvenu OJ, Davydow DS, Kendler KS. 2011. Psychiatric "diseases" versus behavioral disorders and degree of genetic influence. *Psychol Med 41*(1): 33–40.

Blankenburg W. 1971. *Der Verlust der natürlichen Selbstverständlichkeit*. Enke, Stuttgart.

Bleuler E. 1906. Freudsche Mechanismen in der Dementia praecox. In: Bleuler M (Ed.) Beiträge zur Schizophrenielehre der Züricher Universitätsklinik Burghölzli, pp. 22–30. Wissenschaftliche Buchgemeinschaft, Darmstadt. (Reprinted 1979)

Bleuler E. 1911. *Dementia praecox oder die Gruppe der Schizophrenien.* Deuticke, Leizig, Wien.

Bleuler E. 1927. *Das autistisch-undisziplinierte Denken in der Medizin und seine Ueberwindung.* Springer, Berlin.

Blumenbach JF. 1795. *De Generis Humani Varietate Nativa,* 5th ed. Vandenhoeck, Göttingen.

Bobon DP. 1983. *The AMDP-System in Pharmacopsychiatry.* Karger, Basel.

Bock T, Heinz A. 2016. *Psychosen. Ringen um Selbstverständlichkeit.* Psychiatrie-Verlag, Köln.

Boehme R, Deserno L, Gleich T, Katthagen T, Pankow A, Behr J, Buchert R, et al. 2015. Aberrant salience is related to reduced reinforcement learning signals and elevated dopamine synthesis capacity in healthy adults. *J Neurosci 35*: 10103–10111.

Bohman M. 1996. Predisposition to criminality: Swedish adoption studies in retrospect. *Ciba Found Symp 195*: 99–109.

Bohman M, Cloninger CR, Sigvardsson S, von Knorring AL. 1982. Predisposition to petty criminality in Swedish adoptees. *Arch Gen Psychiatry 39*: 1233–1241.

Boorse C. 1976. What a theory of mental health should be. *J Theory Soc Behav 6*: 61–84.

Boorse C. 1977. Health as a theoretical concept. *Philos Sci 44*: 542–573.

Boorse C. 1997. A rebuttal on health. In: Humber JM, Almender RF (Eds.), *What Is Disease?*, pp. 3–134. Humana Press, Totowa, NJ.

Boorse C. 2012. Gesundheit als theoretischer Begriff. In: Schramme T (Ed.), *Krankheitstheorien*, pp. 63–110. Suhrkamp, Berlin.

Braus DF, Wrase J, Grüsser S, Hermann D, Ruf M, Flor H, Mann K, Heinz A. 2001. Alcohol-associated stimuli activate the ventral striatum in abstinent alcoholics. *J Neural Transm 108*: 887–894.

Brodie MS, Bunney EB. 1996. Serotonin potentiates dopamine inhibition of ventral tegmental area neurons in vivo. *J Neurophysiol 76*: 2077–2082.

Brody AL, Saxena S, Schwartz JM, Stoessel PW, Maidment K, Phleps ME, Baxter LR, Jr. 1998. FDG-PET predictors of response to behavioral therapy and pharmacotherapy in obsessive-compulsive disorder. *Psychiatry Res 84*: 1–6.

Büchel C, Dolan R. 2000. Classical fear conditioning in functional neuroimaging. *Curr Opin Neurobiol 10*: 219–223.

Bundesverfassungsgericht (BVerfG). 2011. 2 BvR 882/09. Available at www.bverfg.de /entscheidungen/rk20090622_2bvr088209.html.

Callicott JH, Mattay VS, Verchinski BA, Marenco S, Egan MF, Weinberger DR. 2003. Complexity of prefrontal cortical dysfunction in schizophrenia: More than up or down. *Am J Psychiatry 160*: 2209–2215.

Cantor-Graae E, Selten JP. 2005. Schizophrenia and migration: A meta-analysis and review. *Am J Psychiatry 162*: 12–24.

Carlsson A, Waters N, Carlsson ML. 1999. Neurotransmitter actions in schizophrenia–therapeutic implications. *Biol Psychiatry 46*: 1388–1395.

Carpenter WT, Jr., Strauss JS, Muleh S. 1973. Are there pathognomonic symptoms in schizophrenia? An empiric investigation of Schneider's first-rank symptoms. *Arch Gen Psychiatry 28*: 847–852.

Carter CS, Braver TS, Barch DM, Botvinick MM, Noll D, Cohen JD. 1998. Anterior cingulate, error detection, and the online monitoring of performance. *Science 280*: 747–749.

Cartoni E, Moretta T, Puglisi-Alegra S, Cabib S, Baldassare G. 2015. The relationship between specific Pavlovian instrumental transfer and instrumental reward probability. *Front Psychol 6*: 1697.

Caspi A, McClay J, Moffitt TE, Mill J, Martin J, Craig IW, Taylor A, Poulton R. 2002. Role of genotype in the cycle of violence in maltreated children. *Science 297*: 851–854.

Caspi A, Sugden K, Moffitt TE, Taylor A, Craig IW, Harrington H, McClay J, et al. 2003. Influence of life stress on depression: Moderation by a polymorphism in the 5-HTT gene. *Science 301*(5631): 386–389.

Chamorro AJ, Marcos M, Miron-Canelo JA, Pastor I, Gonzalez-Sarmieto R, Laso FJ. 2012. Association of μ-opioid receptor (OPRM1) gene polymorphism with response to naltrexone in alcohol dependence: A systematic review and meta-analysis. *Addict Biol 17*: 505–512.

Charlet K, Beck A, Jorde A, Wimmer L, Vollstädt-Klein S, Gallinat J, Walter H, et al. 2014. Increased neural activity during high working memory load predicts low relapse risk in alcohol dependence. *Addict Biol 19*: 402–414.

Chen C, Takahashi T, Nakagawa S, Inoue T, Kusumi I. 2015. Reinforcement learning in depression: A review of computational research. *Neurosci Biobehav Rev 55*: 247–267.

Clarke HF, Dalley JW, Crofts HS, Robbins TW, Roberts AC. 2004. Cognitive inflexibility after prefrontal serotonin depletion. *Science 304*: 878–880.

Cloninger CR. 1987a. A systematic method for clinical description and classification of personality variants: A proposal. *Arch Gen Psychiatry 44*: 573–588.

Cloninger CR. 1987b. Neurogenetic adaptive mechanisms in alcoholism. *Science 236*: 410–416.

Cohen JD, Servan-Schreiber D. 1992. Context, cortex, and dopamine: A connectionist approach to behavior and biology in schizophrenia. *Psychol Rev 99*: 45–77.

Conrad K. 1958. *Die beginnende Schizophrenie*. Thieme, Stuttgart.

Cools R, Gibbs SE, Miyakawa A, Jagust W, D'Esposito M. 2008. Working memory capacity predicts dopamine synthesis capacity in the human striatum. *J Neurosci 28*: 1208–1212.

Cools R, Nakamura K, Daw ND. 2011. Serotonin and dopamine: Unifying affective, activational, and decision functions. *Neuropsychopharmacology 36*: 98–113.

Corbit JH, Balleine BW. 2005. Double dissociation of basolateral and central amygdala lesions on the general and outcome-specific forms of Pavlovian-instrumental transfer. *J Neurosci 25*: 962–970.

Corbit LH, Janak PH, Balleine BW. 2007. General and outcome-specific forms of Pavlovian-instrumental transfer: The effect of shifts in motivational state and inactivation of the ventral tegmental area. *Eur J Neurosci 26*: 3141–3149.

Corlett PR, Fletcher PC. 2014. Computational psychiatry: A Rosetta Stone linking the brain to mental illness. *Lancet Psychiatry 1*: 399–402.

Crawford JR, Henry JD. 2004. The positive and negative affect schedule (PANAS): Construct validity, measurement properties and normative data in a large non-clinical sample. *Br J Clin Psychol 43*: 245–265.

Cummings JL. 1993. Frontal-subcortical circuits and human behavior. *Arch Neurol 50*: 873–880.

Czihak G. 1981. *Biologie*. Springer, Berlin.

Daw ND, Kakade S, Dayan P. 2002. Opponent interactions between serotonin and dopamine. *Neural Netw 15*: 603–616.

Daw ND, Niv Y, Dayan P. 2005. Uncertainty-based competition between prefrontal and dorsolateral striatal systems for behavioral control. *Nat Neurosci 8*: 1704–1711.

Daw ND, Gershman SJ, Seymour B, Dayan P, Dolan RJ. 2011. Model-based influences on human choices and striatal prediction errors. *Neuron 69*: 1204–1215.

Demjaha A, Murray RM, McGuire PK, Kapur S, Howes OD. 2012. Dopamine synthesis capacity in patients with treatment-resistant schizophrenia. *Am J Psychiatry 169*: 1203–1210.

den Ouden HE, Ko P, de Lange FP. 2012. How prediction errors shape perception, attention, and motivation. *Front Psychol 3*: 458.

den Ouden HE, Friston KJ, Daw ND, McIntosh AR, Stephan KE. 2009. A dual role for prediction error in associative learning. *Cereb Cortex 19*: 1175–1185.

den Ouden HE, Daunizeau J, Roiser J, Friston KJ, Stephan KE. 2010. Striatal prediction error modulates cortical coupling. *J Neurosci 30*: 3210–3219.

Deserno L, Sterzer P, Wüstenberg T, Heinz A, Schlagenhauf F. 2012. Reduced prefrontal-parietal effective connectivity and working memory deficits in schizophrenia. *J Neurosci 32*: 12–20.

Deserno L, Huys QJ, Boehme R, Buchert R, Heinze HJ, Grace AA, Dolan RJ, et al. 2015a. Ventral striatal dopamine reflects behavioral and neural signatures of model-based control during sequential decision making. *Proc Natl Acad Sci USA 12*: 1595–1600.

Deserno L, Beck A, Huys QJ, Lorenz RC, Buchert R, Buchholz HG, Plotkin M, et al. 2015b. Chronic alcohol intake abolishes the relationship between dopamine synthesis capacity and learning signals in the ventral striatum. *Eur J Neurosci 41*: 47786.

Dettling M, Heinz A, Dufeu P, Rommelspacher H, Gräf KJ, Schmidt LG. 1995. Dopaminergic responsivity in alcoholism: Trait, state, or residual marker? *Am J Psychiatry 152*: 1317–1321.

Dewey SL, Smith GS, Logan J, Alexhoff D, Ding YS, King P, Pappas N, et al. 1995. Serotonergic modulation of striatal dopamine measured with positron emission tomography (PET) and in vivo microdialysis. *J Neurosci 15*: 821–829.

De Wit H. 2008. Impulsivity as a determinant and consequence of drug use: A review of underlying processes. *Addict Biol 14*: 22–31.

Di Chiara G, Bassareo V. 2007. Reward system and addiction: What dopamine does and doesn't do. *Curr Opin Pharmacol 7*: 69–76.

Di Chiara G, Imperato A. 1988. Drugs abused by humans preferentially increase synaptic dopamine concentrations in the mesolimbic system of freely moving rats. *Proc Natl Acad Sci USA 85*: 5274–5278.

Dilthey W. 1924. *Gesammelte Schriften*. Teubner, Leipzig.

Dolan RJ, Dayan P. 2013. Goals and habits in the brain. *Neuron 80*: 312–325.

Doll BB, Bath KG, Daw ND, Frank MJ. 2016. Variability in dopamine genes dissociates model-based and model-free reinforcement learning. *J Neurosci 36*: 1211–1222.

Doudet D, Hommer D, Higley JD, Andreason PJ, Moneman R, Suomi SS, Linnoila M. 1995. Cerebral glucose metabolism, CSF 5-HIAA levels, and aggressive behavior in rhesus monkeys. *Am J Psychiatry 152*: 1782–1787.

Drevets WC. 2000. Neuroimaging studies of mood disorders. *Biol Psychiatry 48*: 813–829.

Drevets WC, Price JL, Simpson JR, Jr., Todd RD, Reich T, Vannier M, Raichle ME. 1997. Subgenual prefrontal cortex abnormalities in mood disorders. *Nature 386*: 824–827.

Drevets WC, Gautier C, Price JC, Kupfer DJ, Kinahan PE, Grace AA, Proce JL, Mathis CA. 2001. Amphetamine-induced dopamine release in human ventral striatum correlates with euphoria. *Biol Psychiatry 49*: 81–96.

Eberl C, Wiers RW, Pawelczack S, Rinck M, Becker ES, Lindenmeyer J. 2014. Implementation of approach bias re-training in alcoholism—how many sessions are needed? *Alcohol Clin Exp Res 38*: 587–594.

Edwards G. 1990. Withdrawal symptoms and alcohol dependence: Fruitful mysteries. *Br J Addict 85*: 447–461.

Egan MF, Goldberg TE, Kolachana BS, Callicott JH, Mazzanti CM, Straub RE, Goldman D, Weinberger DR. 2001. Effect of COMT Val108/158 Met genotype on frontal lobe function and risk for schizophrenia. *Proc Natl Acad Sci USA 98*: 6917–6922.

Egerton A, Ahmad R, Hirani E, Grasby PM. 2008. Modulation of striatal dopamine release by 5–HT2A and 5–HT2C receptor antagonists: [11C] raclorpide PET studies in the rat. *Psychopharmacology (Berl) 200*: 487–496.

Enoch MA, Shen PH, Xu K, Hodgkinson C, Goldman D. 2006. Using ancestry-informative markers to define populations and detect population stratification. *J Psychopharmacol 20*(Suppl): 19–26.

Ersche KD, Gillan CM, Jones PS, Williams GB, Ward LH, Luijten M, de Wit S, et al. 2016. Carrots and sticks fail to change behavior in cocaine addiction. *Science 352*: 1468–1471.

Ester EF, Srague TC, Serences JT. 2015. Parietal and prefrontal cortex encode stimulus-specific mnemonic representations during visual working memory. *Neuron 87*: 893–905.

Everitt BJ, Robbins TW. 2005. Neural systems of reinforcement for drug addiction: From actions to habits to compulsion. *NatNeurosci 8*: 1481–1489.

Everitt BJ, Robbins TW. 2016. Drug addiction: Updating actions to habits to compulsions ten years on. *Annu Rev Psychol 67*: 23–50.

Eysenck HJ. 1967. *The Biological Basis of Personality*. Thomas, Springfield, IL.

Faccidomo S, Bannai M, Miczek KA. 2008. Escalated aggression after alcohol drinking in male mice: Dorsal raphé and prefrontal cortex serotonin and 5-HT(1B) receptors. *Neuropsychopharmacology 33*: 2888–2899.

Feeser M, Schlagenhauf F, Sterzer P, Park S, Stoy M, Gutwinski S, Dalanay U, et al. 2013. Context insensitivity during positive and negative emotional expectancy in depression assessed with functional magnetic resonance imaging. *Psychiatry Res 212*: 28–35.

Fiorillo CD, Tobler PN, Schultz W. 2003. Discrete coding of reward probability and uncertainty by dopamine neurons. *Science 299*: 1898–1902.

Flagel SB, Watson SJ, Akil H, Robinson TE. 2008. Individual differences in the attribution of incentive salience to a reward-related cue: Influence on cocaine sensitization. *Behav Brain Res 186*: 48–56.

Flagel SB, Clark JJ, Robinson TE, Mayo L, Czuj A, Willuhn I, Akers CA, et al. 2011. A selective role for dopamine in reward learning. *Nature 469*: 53–57.

Flandreau EI, Ressler KJ, Owens MJ, Nemeroff CB. 2012. Chronic overexpression of corticotropin-releasing factor from the central amygdala produces HPA axis hyperactivity and behavioral anxiety associated with gene-expression changes in the hippocampus and paraventricular nucleus of the hypothalamus. *Psychoneuroendocrinology 37*: 27–38.

Fleck L. 1979. *The Genesis and Development of a Scientific Fact*. University of Chicago Press, Chicago.

Foddy B, Savulescu J. 2007. Addixtion is not an affliction: addictive desires are merely pleasure-oriented desires. *Am J Bioeth 7*: 29–32.

Frances A. 2013. *Saving Normal: An Insider's Revolt against Out-of-Control Psychiatric Diagnosis, DSM-5, Big Pharma, and the Medicalization of Ordinary Life*. HarperCollins, New York.

Frances A, Raven M. 2013. Two views on the DSM-5: The need for caution in diagnosing and treating mental disorders. [online] *Am Fam Physician* 88.

Frank M. 1991. *Selbstbewusstsein und Selbsterkenntnis*. Reclam, Stuttgart.

Frank M. 2012. *Ansichten der Subjektivität*. Suhrkamp, Berlin.

Freud S. 1900. *Die Traumdeutung*. In: Freud S. *Gesammelte Werke*, Vol. II/III, pp. 1–724. Fischer, Frankfurt /M. (Reprint 1977).

Freud S. 1906. *Drei Abhandlungen zur Sexualtheorie*. Fischer, Frankfurt/M. (Reprint 1981).

Freud S. 1911a. *Psychoanalytische Bemerkungen über einen autobiografisch beschriebenen Fall von Paranoia (Dementia paranoides)*. In: Freud S. *Gesammelte Werke*, Vol. VIII, pp. 239–320. Fischer, Frankfurt /M. (Reprint 1977).

Freud S. 1911b. *Formulierungen über die zwei Prinzipien psychischen Geschehens*. In: Freud S. *Gesammelte Werke*, Vol. VIII, pp. 230–238. Fischer, Frankfurt/M. (Reprint 1977).

Freud S. 1913 *Totem und Tabu. Einige Übereinstimmungen im Seelenleben der Wilden und der Neurotiker.* In: Freud S. *Gesammelte Werke,* Vol. IX, pp. 1–207. Fischer, Frankfurt/M. (Reprint 1977)

Freud S. 1938. Abriss der Psychoanalyse. In: Freud S. *Gesammelte Werke,* Vol. XVII, pp. 63–138. Fischer, Frankfurt/M. (Reprint 1977)

Friedel E, Schlagenhauf F, Sterzer P, Park SQ, Bermpohl F, Ströhle A, Stoy M, et al. 2009. 5-HTT genotype effect on prefrontal-amygdala coupling differs between major depression and controls. *Psychopharm (Berlin) 205:* 261–271.

Friedel E, Koch SP, Wendt J, Heinz A, Deserno L, Schlagenhauf F. 2014. Devaluation and sequential decisions: Linking goal-directed and model-based behavior. *Front Hum Neurosci 8:* 587.

Friedel E, Schlagenhauf F, Beck A, Dolan RJ, Huys QJ, Rapp MA, Heinz A. 2015. The effects of life stress and neural learning signals on fluid intelligence. *Eur Arch Psychiatry Clin Neurosci 265:* 35–43.

Friston KJ, Stephan KE, Montague R, Dolan RJ. 2014. Computational psychiatry: The brain as a phantastic organ. *Lancet Psychiatry 1:* 148–158.

Frith CD. 1992. *The Cognitive Neuropsychology of Schizophrenia.* Erlbaum, Hillsdale, NJ.

Froehlich JC, Zink RW, Li TK, Christian JC. 2000. Analysis of heritability of hormonal responses to alcohol in twins: Beta-endorphin as a potential biomarker of genetic risk for alcoholism. *Alcohol Clin Exp Res 24:* 265–277.

Fuchs H, Nagel J, Hauber W. 2005. Effects of physiological and pharmacological stimuli on dopamine release in the rat globus pallidus. *Neurochem Int 47:* 474–481.

Fuallana MA, Mataix-Cols D, Caseras X, Alonso P, Manuel Menchón J, Vallejo J, Torrubia R. 2004. High sensitivity to punishment and low impulsivity in obsessive-compulsive patients with hoarding symptoms. *Psychiatry Res 129*(1): 21–27.

Fuster JM. 2001. The prefrontal cortex—an update. Time is of the essence. *Neuron 30:* 319–333.

Gallagher S. 2000. Philosophical conceptions of the Self. Implications for cognitive science. *Trends Cogn Sci 4:* 14–21.

Gallagher S. 2004. Neurocognitive models of schizophrenia. A neurophenomenological critique. *Psychopathology 37:* 8–19.

Garbusow M, Schad DJ, Sommer C, Jünger E, Sebold M, Friedel E, Wendt J, et al. 2014. Pavlovian-to-instrumental transfer in alcohol dependence: A pilot study. *Neuropsychobiology 70:* 111–121.

Garbusow M, Schad DJ, Sebold M, Friedel E, Bernhardt N, Koch SP, Steinacher B, et al. 2016a. Pavlovian-to-instrumental transfer effects in the nucleus accumbens relate to relapse in alcohol dependence. *Addict Biol 21*: 719–731.

Garbusow M, Schad DJ, Sebold M, Nebe S, Sommer C, Zimmermann US, Smolka MN, Schlagenhauf F, Huys QM, Rapp MA, Heinz A. 2016b. Neruobiological correlates of alcohol-related Pavlovian-to-instrumental transfer and relapse behavior in alcohol dependence: the LeAD study. *Intrins Act 4* (Suppl. 2): A18.34.

Gianoulakis C, Krishnan B, Thavundayil J. 1996. Enhanced sensitivity of pituitary beta-endorphin to ethanol in subjects at high risk of alcoholism. *Arch Gen Psychiatry 53*: 250–257.

Gillan CM, Otto AR, Phelps EA, Daw ND. 2015. Model-based learning protects against forming habits. *Cogn Affect Behav Neurosci 15*: 523–536.

Glascher J, Daw ND, Dayan P, O'Doherty JP. 2010. States versus rewards: Dissociable neural prediction error signals underlying model-based and model-free reinforcement learning. *Neuron 66*: 585–595.

Godemann F, Scharbowska A, Nettebusch B, Heinz A, Ströhle A. 2006. The impact of cognitions on the development of panic and somatoforme disorders: A prospective study in patients with vestibular neuritis. *Psychol Med 36*: 99–108.

Godemann F, Schuller J, Uhlemann H, Budde A, Heinz A, Ströhle A, Hauth I. 2009. Psychodynamic vulnerability factors in the development of panic disorders: A prospective trial in patients after vestibular neuritis. *Psychopathology 42*: 99–107.

Goldman-Rakic PS. 1996. Regional and cellular fractionation of working memory. *Proc Natl Acad Sci USA 93*: 13473–13480.

Goldstein RZ, Volkow ND. 2011. Dysfunction of the prefrontal cortex in addiction: Neuroimaging findings and clinical implications. *Nat Rev Neurosci 12*: 652–669.

Goldstein RZ, Woicik PA, Moeller SJ, Telang F, Jayne M, Wong C, Wang GJ, et al. 2010. Liking and wanting of drug and non-drug rewards in active cocaine users: The STRAP-R questionnaire. *J Psychopharmacol 24*: 257–266.

Goto Y, Grace AA. 2008. Limbic and cortical information processing in the nucleus accumbens. *Trends Neurosci 31*: 552–558.

Grace AA. 1991. Phasic versus tonic dopamine release and the modulation of dopamine system responsivity: A hypothesis for the etiology of schizophrenia. *Neuroscience 41*: 1–24.

Grace AA. 2016. Dysregulation of the dopamine system in the pathophysiology of schizophrenia. *Nat Rev Neurosci 17*: 524–532.

Gray JA. 1982. *The Neuropsychology of Anxiety. An Inquiry into the Function of the Septo-Hippocampal System.* Oxford University Press, New York.

Grubaugh AL, Zinzow HM, Paul L, Egede LE, Frueh BH. 2011. Trauma exposure and posttraumatic stress disorder in adults with severe mental illness: A critical review. *Clin Psychol Rev 31*: 883–899.

Gründer G, Vernaleken I, Müller MJ, Davids E, Heydari N, Buchholz HG, Bartenstein P, et al. 2003. Subchronic haloperidol downregulates dopamine synthesis capacity in the brain of schizophrenic patients in vivo. *Neuropsychopharmacology 28*: 787–794.

Grüsser SM, Wrase J, Klein S, Hermann D, Smolka MN, Ruf M, Weber-Fahr W, et al. 2004. Cue-induced activation of the striatum and medial prefrontal cortex is associated with subsequent relapse in abstinent alcoholics. *Psychopharmacology (Berl) 175*: 296–302.

Gryglewski G, Lanzenberger R, Kranz GS, Cumming P. 2014. Meta-analysis of molecular imaging of serotonin transporters in major depression. *J Cereb Blood Flow Metab 34*(7): 1096–1103.

Gulbins E, Palmada M, Reichel M, Lüth A, Böhmer C, Amato D, Müller CP, et al. 2013. Acid sphingomyelinase-ceramide system mediates effects of antidepressant drugs. *Nat Med 19*: 934–938.

Haber SN, Fudge JL, McFarland NR. 2000. Striatonigrostriatal pathways in primates form an ascending spiral from the shell to the dorsolateral striatum. *J Neurosci 20*: 2369–2382.

Haeckel E. 1866. *Generelle Morphologie der Organismen. Bd. 2: Allgemeine Entwickelungsgeschichte der Organismen.* Georg Reimer, Berlin.

Hafner S. 1994. *Der Verrat. 1918/1919 – als Deutschland wurde, was es ist.* Verlag 1900, Berlin.

Hägele C, Schlagenhauf F, Rapp M, Sterzer P, Beck A, Bermpohl F, Stoy M, et al. 2015. Dimensional psychiatry: Reward dysfunction and depressive mood across psychiatric disorders. *Psychopharmacology (Berl) 232*: 331–341.

Hägele C, Friedel E, Schlagenhauf F, Sterzer P, Beck A, Bermpohl F, Stoy M, et al. 2016. Affective responses across psychiatric disorders—A dimensional approach. *Neurosci Lett 623*: 71–78.

Hamani C, Mayberg H, Stone S, Laxton A, Haber S, Lozano AM. 2011. The subcallosal cingulated gyrus in the context of major depression. *Biol Psychiatry 69*: 301–308.

Hampton AN, Bossaerts P, O'Doherty JP. 2006. The role of the ventromedial prefrontal cortex in abstract state-based inference during decision-making in humans. *J Neurosci 26*: 8360–8367.

Hariri AR, Mattay VS, Tessitore A, Kolachana B, Fera F, Goldman D, Egan MF, Weinberger DR. 2002. Serotonin transporter genetic variation and the response of the human amygdala. *Science 297*(5580): 400–403.

Hauser M, Knoblich G, Repp BH, Lautenschlager M, Gallinat J, Heinz A, Voss M. 2011. Altered sense of agency in schizophrenia and the putative psychotic prodrome. *Psychiatry Res 186*: 170–176.

He H, Yu Q, Du Y, Vergara V, Victor TA, Drevets WC, Savitz JB, et al. 2016. Resting-state functional network connectivity in prefrontal regions differs between unmedicated patients with bipolar and major depressive disorder. *J Affect Disord 190*: 483–493.

Heckers S. 2001. Neuroimaging studies of the hippocampus in schizophrenia. *Hippocampus 11*: 520–528.

Heckers S, Konradi C. 2015. GABAergic mechanisms of hippocampus hyperactivity in schizophrenia. *Schizophr Res 167*: 4–11.

Heidegger M. 2006. *Sein und Zeit*. Niemeyer, Tübingen.

Heim C, Mietzko T, Purselle D, Musselman DL, Nemeroff CB. 2008. The dexamethasone/corticotropin-releasing factor test in men with major depression: Role of childhood trauma. *Biol Psychiatry 63*: 398–405.

Heinz A. 1998. Colonial perspectives in the construction of the psychotic patient as primitive man. *Crit Anthropol 18*: 421–444.

Heinz A. 1999. Neurobiological and anthropological aspects of compulsions and ritual. *Pharmacopsychiatry 32*: 223–229.

Heinz A. 2002a. Dopaminergic dysfunction in alcoholism and schizophrenia—psychopathological and behavioral correlates. *Eur Psychiatry 17*: 9–16.

Heinz A. 2002b. *Evolutionäre und anthropologische Modelle in der Schizophrenieforschung*. Verlag für Bildung und Wissenschaft, Berlin.

Heinz A. 2014. *Der Begriff der psychischen Krankheit*. Suhrkamp, Berlin.

Heinz A, Heinze M. 1999. From pleasure to anhedonia—forbidden desires and the construction of schizophrenia. *Theory Psychol 9*: 47–65.

Heinz A, Schlagenhauf F. 2010. Dopaminergic dysfunction in schizophrenia: Salience attribution revisited. *Schizophr Bull 36*: 472–485.

Heinz A, Weinberger DR. 2000. *Schizophrenia: The Neurodevelopmental Hypothesis*, pp. 89–104. Current Concepts in Psychiatry (Psychiatrie der Gegenwart). Springer, Berlin.

Heinz A, Schmidt LG, Reischies FM. 1994. Anhedonia in schizophrenic, depressed, or alcohol-dependent patients—neurobiological correlates. *Pharmacopsychiatry 27*(Suppl 1): 7–10.

Heinz A, Przuntek H, Winterer G, Pietzcker A. 1995a. Clinical aspects and follow-up of dopamine-induced psychoses in continuous dopaminergic therapy and their implications for the dopamine hypothesis of schizophrenic symptoms. *Nervenarzt* 66: 662–669.

Heinz A, Lichtenberg-Kraag B, Sällström Baum S, Gräf K, Krüger F, Dettling M, Rommelspacher H. 1995b. Evidence for prolonged recovery of dopaminergic transmission in alcoholics with poor treatment outcome. *J Neural Transm 102*: 149–158.

Heinz A, Dufeu P, Kuhn S, Dettling M, Gräf K, Kürten I, Rommelspacher H, Schmidt LG. 1996a. Psychopathological and behavioral correlates of dopaminergic sensitivity in alcohol-dependent patients. *Arch Gen Psychiatry 53*: 1123–1128.

Heinz A, Bauer M, Kuhn S, Krüger F, Gräf KJ, Rommelspacher H, Schmidt LG. 1996b. Long-term observation of the hypothalamic-pituitary-thyroid (HPT) axis in alcohol-dependent patients. *Acta Psychiatr Scand 93*(6): 470–476.

Heinz A, Higley JD, Gorey JG, Saunders RC, Jones DW, Hommer D, Zajicek K, et al. 1998a. In vivo association between alcohol intoxication, aggression, and serotonin transporter availability in nonhuman primates. *Am J Psychiatry 155*: 1023–1028.

Heinz A, Knable MB, Coppola R, Gorey JG, Jones DW, Lee KS, Weinberger DR. 1998b. Psychomotor slowing, negative symptoms and dopamine receptor availability—an IBZM SPECT study in neuroleptic-treated and drug-free schizophrenic patients. *Schizophr Res 31*: 19–26.

Heinz A, Ragan P, Jones DW, Hommer D, Williams W, Knable MB, Gorey J, et al. 1998c. Reduced serotonin transporters in alcoholism. *Am J Psychiatry 155*: 1544–1549.

Heinz A, Knable MB, Wolf SS, Jones DW, Gorey JG, Hyde TM, Weinberger DR. 1998d. Tourette's syndrome: [l-123]beta-CIT SPECT correlates of vocal tic severity. *Neurology 51*: 1069–1074.

Heinz A, Saunders RC, Kolachana BS, Jones DW, Gorey JG, Bachevalier J, Weinberger DR. 1999. Striatal dopamine receptors and transporters in monkeys with neonatal temporal limbic damage. *Synapse 32*: 71–79.

Heinz A, Jones DW, Mazzanti C, Goldman D, Ragan P, Hommer D, Linnoila M, Weinberger DR. 2000a. A relationship between serotonin transporter genotype and in vivo expression and alcohol neurotoxicity. *Biol Psychiatry 47*: 643–649.

Heinz A, Goldman D, Jones DW, Palmour R, Hommer D, Gorey JG, Lee KS, et al. 2000b. Genetic effects on striatal dopamine transporter availability in human striatum. *Neuropsychopharmacology 22*: 133–139.

Heinz A, Mann K, Weinberger DR, Goldman D. 2001. Serotonergic dysfunction, negative mood states, and response to alcohol. *Alcohol Clin Exp Res 25*: 487–495.

Heinz A, Jones DW, Bissette G, Hommer D, Ragan P, Knable M, Wellek S, et al. 2002. Relationship between cortisol and serotonin metabolites and transporters in alcoholism. *Pharmacopsychiatry 35*: 127–134.

Heinz A, Löber S, Georgi A, Wrase J, Hermann D, Rey ER, Wellek S, Mann K. 2003a. Reward craving and withdrawal relief craving: Assessment of different motivational pathways to alcohol intake. *Alcohol Alcohol 38*: 35–39.

Heinz A, Jones DW, Gorey JG, Bennet A, Suomi SJ, Weinberger DR, Higley JD. 2003b. Serotonin transporter availability correlates with alcohol intake in non-human primates. *Mol Psychiatry 8*: 231–234.

Heinz A, Romero B, Weinberger DR. 2004a. Functional mapping with single-photon-emission computed tomography and positron emission tomography. In: Lawrie S, Johnstone E, Weinberger DR (Eds.), *Schizophrenia—from Neuroimaging to Neuroscience*, pp. 167–211. Oxford University Press, Oxford.

Heinz A, Siessmeier T, Wrase J, Hermann D, Klein S, Grüsser SM, Flor H, et al. 2004b. Correlation between dopamine D(2) receptors in the ventral striatum and central processing of alcohol cues and craving. *Am J Psychiatry 161*: 1783–1789.

Heinz A, Braus DF, Smolka MN, Wrase J, Puls I, Hermann D, Klein S, et al. 2005a. Amygdala-prefrontal coupling depends on a genetic variation of the serotonin transporter. *Nat Neurosci 8*: 20–21.

Heinz A, Reimold M, Wrase J, Hermann D, Croissant B, Mundle G, Dohmen BM, et al. 2005b. Correlation of stable elevations in mu-opiate receptor availability in detoxified alcoholic patients with alcohol craving: A positron emission tomography study using carbon 11-labeled carfentanil. *Arch Gen Psychiatry 62*: 5764.

Heinz A, Siessmeier T, Wrase J, Buchholz HG, Gründer G, Kumakura Y, Cumming P, et al. 2005c. Correlation of alcohol craving with striatal dopamine synthesis capacity and D2/3 receptor availability: A combined [18F]DOPA and [18F]DMFP PET study in detoxified alcoholic patients. *Am J Psychiatry 162*: 1515–1520.

Heinz A, Smolka MN, Braus DF, Wrase J, Beck A, Flor H, Mann K, et al. 2007. Serotonin transporter genotype (5-HTTLPR): Effects of neutral and undefined conditions on amygdala activation. *Biol Psychiatry 61*: 1011–1014.

Heinz A, Beck A, Wrase J, Mohr J, Obermayer K, Gallinat J, Puls I. 2009. Neurotransmitter systems in alcohol dependence. *Pharmacopsychiatry 42*(Suppl 1): S95–S101.

Heinz AJ, Beck A, Meyer-Lindenberg A, Sterzer P, Heinz A. 2011. Cognitive and neurobiological mechanisms of alcohol-related aggression. *Nat Rev Neurosci 12*: 400–413.

Heinz A, Bermpohl F, Frank M. 2012a. Construction and interpretation of self-related function and dysfunction in intercultural psychiatry. *Eur Psychiatry 2*: 32–43.

Heinz A, Batra A, Scherbaum N, Gouzoulis-Mayfrank E. 2012b. *Neurobiologie der Abhängigkeit*. Kohlhammer, Stuttgart.

Heinz A, Deserno L, Reininghaus U. 2013. Urbanicity, social adversity and psychosis. *World Psychiatry 12*: 187–197.

Heinz A, Müller DJ, Krach S, Cabanis M, Kluge UP. 2014. The uncanny return of the race concept. *Front Hum Neurosci 8*: 836.

Heinz A, Voss M, Lawrie SM, Mishara A, Bauer M, Gallinat J, Juckel G, et al. 2016a. Shall we really say good-bye to first rank symptoms? *Eur Psychiatry 37*: 8–13.

Heinz A, Deserno L, Zimmermann US, Smolka MN, Beck A, Schlagenhauf F. 2016b. Targeted intervention: Computational approaches to elucidate and predict relapse in alcoholism. *Neuroimage* [epub ahead of print] pii:S1053–8119(16)30369-X.

Helmchen H, Hippius H. 1967. Depression syndrome in the course of neuroleptic therapy. *Nervenarzt 38*: 455–458.

Helzer JE, Robins LN, Taylor JR, Carey K, Miller RH, Combs-Orme T, Farmer A. 1985. The extent of long-term moderate drinking among alcoholics discharged from medical and psychiatric treatment facilities. *N Engl J Med 312*: 1678–1682.

Hermann D, Hirth N, Reimold M, Batra A, Smolka MN, Hoffmann S, Kiefer F, et al. 2017. Low μ-opioid receptor status in alcohol dependence identified by combining positron emission tomography and post-mortem brain analysis. *Neuropsychopharmacology 42*: 606–614. doi:10.1038/npp-2016-145.

Higley JD, Suomi SS, Linnoila M. 1991. CSF monoamine metabolite concentrations vary according to age, rearing, and sex, and are influenced by the stressor of social separation in rhesus monkeys. *Psychopharmacology (Berl) 103*: 551–556.

Higley JD, Suomi SS, Linnoila M. 1992a. A longitudinal assessment of CSF monoamine metabolite and plasma cortisol concentrations in young rhesus monkeys. *Biol Psychiatry 32*: 127–145.

Higley JD, Mehlman PT, Taub DM, Higley SB, Suomi SS, Linnoila M, Vickers JH. 1992b. Cerebrospinal fluid monoamine and adrenal correlates of aggression in free-ranging rhesus monkeys. *Arch Gen Psychiatry 49*: 436–441.

Higley JD, Suomi SJ, Linnoila M. 1996a. A nonhuman primate model of type II alcoholism? Part 2. Diminished social competence and excessive aggression correlates with low cerebrospinal fluid 5-hydroxyindoleacetic acid concentrations. *Alcohol Clin Exp Res 20*: 643–650.

Higley JD, Suomi SJ, Linnoila M. 1996b. A nonhuman primate model of type II excessive alcohol consumption? Part 1. Low cerebrospinal fluid 5-hydroxyindoleacetic acid concentrations and diminished social competence correlate with excessive alcohol consumption. *Alcohol Clin Exp Res 20*: 629–642.

Hinckers A, Laucht M, Heinz A, Schmidt MH. 2005. Risikoverhalten und Alkoholkonsum im Jugendalter. *Z Kinder Jugendpsychiatr Psychother 33*: 273–282.

Hinckers AS, Laucht M, Schmidt MH, Mann KF, Schumann G, Schuckit MA, Heinz A. 2006. Low level of response to alcohol is associated with serotonin transporter genotype and high alcohol intake in adolescents. *Biol Psychiatry 60*: 282–287.

Hinson RE, Siegel S. 1982. Nonpharmacological bases of drug tolerance and dependence. *J Psychosom Res 26*: 495–503.

Hirth N, Meinhardt MW, Noori HR, Torres-Ramirez O, Uhrig S, Broccoli L, Vengeliene V, et al. 2016. Convergent evidence from alcohol-dependent humans and rats for a hyperdopaminergic state in protracted abstinence. *Proc Natl Acad Sci USA 113*: 3024–3029.

Ho BC, Andreasen NC, Ziebell S, Pierson R, Magnotta V. 2011. Long-term antipsychotic treatment and brain volumes: A longitudinal study of first episode schizophrenia. *Arch Gen Psychiatry 68*: 128–137.

Holtzheimer PE, Mayberg H. 2011. Stuck in a rut: Rethinking depression and its treatment. *Trends Neurosci 34*: 1–9.

Homan P, Neumeister A, Nugent AC, Charney DS, Drevets WC, Hasler G. 2015. Serotonin versus catecholamine deficiency: Behavioral and neural effects of experimental depletion in remitted depression. *Transl Psychiatry 5*: e532.

Howes OD, Kambeitz J, Kim E, Stahl D, Slifstein M, Abi-Dargham A, Kapur S. 2012. The nature of dopamine dysfunction in schizophrenia and what this means for treatment. *Arch Gen Psychiatry 69*: 776–786.

Huys QJ, Cools R, Gölzer M, Friedel E, Heinz A, Dolan RJ, Dayan P. 2011. Disentangling the roles of approach, activation and valence in instrumental and Pavlovian responding. *PLOS Comput Biol 7*: e1002028.

Huys QJM, Deserno L, Obermayer K, Schlagenhauf F, Heinz A. 2016a. Model-free temporal-difference learning and dopamine in alcohol dependence: Examining concepts from theory and animals in human imaging. *Biological Psychiatry: CNNI. 1*: 401–410.

Huys QJ, Gölzer M, Friedel E, Heinz A, Cools R, Dayan P, Dolan RJ. 2016b. The specificity of Pavlovian regulation is associated with recovery from depression. *Psychol Med 46*: 1027–1035.

Hyman SE. 2005. Addiction: A disease of learning and memory. *Am J Psychiatry 162*: 1414–1422.

Imber S, Schultz E, Funderburk F, Allen R, Flamer R. 1976. The fate of the untreated alcoholic. Toward a natural history of the disorder. *J Nerv Ment Dis 162*: 238–247.

Ingwood B, Gerson LP. 1994. *The Epicurus Reader. Selected Writings and Testimonia.* Hackett, Indianapolis.

Insel T, Cuthbert B, Garvey M, Heinssen R, Pine DS, Quinn K, Sanislow C, Wang P. 2010. Research domain criteria (RDoC): Toward a new classification framework for research on mental disorders. *Am J Psychiatry 167*: 748–751.

International Schizophrenia Consortium, Purcell SM, Wray NR, Stone JL, Visscher PM, O'Donovan MC, Sullivan PF, Sklar P. 2009. Common polygenic variation contributes to risk of schizophrenia and bipolar disorder. *Nature 460*: 748–752.

Jablensky A, Sartorius N. 2008. What did the WHO studies really find? *Schizophr Bull 34*: 253–255.

Jackson JH. 1884. The Croonian Lectures on Evolution and Dissolution of the Nervous System. *Lancet 123*: 739–744.

Jardri R, Duverne S, Litvinova AS, Deneve S. 2017. Experimental evidence for circular inference in schizophrenia. *Nat Commun 8*: 14218.

Jaspers K. 1946. Allgemeine Psychopathologie. Springer, Berlin, Heidelberg.

Johnson EO, van den Bree MB, Pickens RW. 1996. Indicators of genetic and environmental influence in alcohol-dependent individuals. *Alcohol Clin Exp Res 20*: 67–74.

Jonas DE, Amick HR, Feltner C, Wines R, Shanahan E, Rowe CJ, Garbutt JC. 2014. Genetic polymorphisms and response to medications for alcohol use disorders: A systematic review and meta-analysis. *Pharmacogenomics 15*: 1687–1700.

Juckel G, Mendlin A, Jacobs BL. 1999. Electric stimulation of rat medial prefrontal cortex enhances forebrain serotonin output: Implications for electro convulsive therapy and transcranial magnetic stimulation in depression. *Neuropsychopharmacology 21*: 391–398.

Juckel G, Schlagenhauf F, Koslowski M, Wüstenberg T, Villringer A, Knutson B, Wrase J, Heinz A. 2006. Dysfunction of ventral striatal reward prediction in schizophrenia. *Neuroimage 29*: 409–416.

Jung CG. 1907. Über die Psychologie der Dementia praecox. In: Jung CG. *Psychogenese der Geisteskrankheiten. Gesammelte Werke*, Vol. 3, pp. 33–288. Rascher, Zürich. (Reprint 1968)

Kahn E. 1919. Psychopathie und Revolution. *Münchner Med Wochenzeitschr 66*: 968–969.

Kalivas PW, Volkow ND. 2005. The neural basis of addiction: A pathology of motivation and choice. *Am J Psychiatry 162*: 1403–1413.

Kanfer FH, Saslow G. 1965. Behavioral analysis: An alternative to diagnostic classification. *Arch Gen Psychiatry 12*: 529–538.

Kant I. 1983. Anthropologie in pragmatischer Hinsicht. Reclam, Stuttgart.

Kapur S. 2003. Psychosis as a state of aberrant salience: A framework linking biology, phenomenology, and pharmacology in schizophrenia. *Am J Psychiatry 160*: 13–23.

Karg K, Burmeister M, Shedden K, Sen S. 2011. The serotonin transporter promoter variant (5-HTTLPR), stress, and depression meta-analysis revisited: Evidence of genetic moderation. *Arch Gen Psychiatry 68*(5): 444–454.

Kaufmann C, Beucke JC, Preuße F, Endrass T, Schlagenhauf F, Heinz A, Juckel G, Kathmann N. 2013. Medial prefrontal brain activation to anticipated reward and loss in obsessive-compulsive disorder. *Neuroimage Clin 17*(2): 212–220.

Kegeles LS, Abi-Dargham A, Zea-Ponce Y, Rodenhiser-Hill J, Mann JJ, Van Heertum RL, Cooper TB, et al. 2000. Modulation of amphetamine-induced striatal dopamine release by ketamine in humans: Implications for schizophrenia. *Biol Psychiatry 48*: 627–640.

Kegeles LS, Abi-Dargham A, Frankle WG, Gil R, Cooper TB, Slifstein M, Hwang DR, et al. 2010. Increased synaptic dopamine function in associative regions of the striatum in schizophrenia. *Arch Gen Psychiatry 67*: 231–239.

Kienast T, Hariri AR, Schlagenhauf F, Wrase J, Sterzer P, Buchholz HG, Smolka MN, et al. 2008. Dopamine in amygdala gates limbic processing of aversive stimuli in humans. *Nat Neurosci 11*: 1381–1382.

Kirsch I, Deacon B, Huedo-Medina TB, Scoboria A, Moore TJ, Johnson BT. 2008. Initial Severity and antidepressant benefits: A meta-analysis of data submitted to the Food and Drug Administration. *PLoS Med 5*: 260–266.

Knutson B, Adams CM, Fong GW, Hommer D. 2001. Anticipation of increasing monetary reward selectively recruits nucleus accumbens. *J Neurosci 21*: RC159.

Kobiella A, Reimold M, Ulshöfer DE, Ikonomidou VN, Vollmert C, Vollstädt-Klein S, Rietschel M, et al. 2011. How the serotonin transporter 5-HTTLPR polymorphism influences amygdale function: The role of in vivo serotonin transporter expression and amygdale function. *Transl Psychiatry 1*: e37.

Köhler W. 1925. *The Mentality of Apes*. Kegan Paul, Trench, Trubner, London.

Koob GF, Le Moal M. 1997. Drug abuse: Hedonic homeostatic dysregulation. *Science 278*: 52–58.

Koob GF, Le Moal M. 2006. *Neurobiology of Addiction*. Elsevier Academic, Cambridge, MA.

Kraepelin E. 1913. *Psychiatrie. Ein Lehrbuch für Studierende und Ärzte*. Barth Verlag, Leipzig.

Kraepelin E. 1916. *Einführung in die Psychiatrische Klinik*. Barth Verlag, Leipzig.

Kraepelin E. 1919. Psychiatrische Randbemerkungen zur Zeitgeschichte. *Süddeutsche Monatshefte xvi:* 171–183.

Kraepelin E. 1920. Die Erscheinungsformen des Irreseins. *Z Gesamte Neurol Psychiatr 62*: 1–29.

Kraepelin E. 1921. Über Entwurzelung. *Zeitschr ges Neurol Psychiatr 63*: 1–8.

Krystal JH, Staley J, Mason G, Petrakis IL, Kaufman J, Harris RA, Gelernter J, Lappalainen J. 2006. Gamma-aminobutyric acid type A receptors and alcoholism: Intoxication, dependence, vulnerability, and treatment. *Arch Gen Psychiatry 63*: 957–968.

Kuhn T. 1962. *The Structure of Scientific Revolutions*. University of Chicago Press, Chicago.

Kühn S, Gallinat J. 2011. Common biology of craving across legal and illegal drugs— a quantitative meta-analysis of cue-reactivity brain response. *Eur J Neurosci 3*: 1318–1326.

Kumakura Y, Cumming P, Vernaleken I, Buchholz HG, Siessmeier T, Heinz A, Kienast T, et al. 2007. Elevated [18F]fluorodopamine turnover in brain of patients with schizophrenia: An [18F]fluorodopa/positron emission tomography study. *J Neurosci 27*: 8080–8087.

Kumar P, Waiter G, Ahearn T, Milders M, Reid I, Steele JD. 2008. Abnormal temporal difference reward-learning signals in major depression. *Brain 131*: 2084–2093.

Lacan J. 1986. *The Ethics of Psychoanalysis*. Editions du Seuil, Paris.

Laine TP, Ahonen A, Räsänen P, Tiihonen J. 1999. Dopamine transporter availability and depressive symptoms during alcohol withdrawal. *Psychiatry Res 90*: 153–157.

Lang P. 1995. The emotion probe. Studies of motivation and attention. *Am Psychol 50*: 372–385.

Laruelle M, Abi-Dargham A, van Dyck CH, Gil R, D'Souza CD, Erdos J, McCance E, et al. 1996. Single photon emission computerized tomography imaging of amphetamine-induced dopamine release in drug-free schizophrenic subjects. *Proc Natl Acad Sci USA 93*: 9235–9240.

Lee SW, Shimojo S, O'Doherty JP. 2014. Neural computations underlying arbitration between model-based and model-free learning. *Neuron 81*: 687–699.

Lesch KP, Bengel D, Heils A, Sabol SZ, Greenberg BD, Petri S, Benjamin J, et al. 1996. Association of anxiety-related traits with a polymorphism in the serotonin transporter regulatory region. *Science 274*: 1527–1531.

Lewontin RC. 1972. The apportionment of human diversity. *J Evol Biol 6*: 381–398.

Lipska BK, Weinberger DR. 2002. A neurodevelopmental model of schizophrenia: Neonatal disconnection of the hippocampus. *Neurotox Res 4*: 469–475.

Lipska BK, Jaskiw GE, Weinberger DR. 1994. The effects of combined prefrontal cortical and hippocampal damage on dopamine-related behaviors in rats. *Pharmacol Biochem Behav 48*: 1053–1057.

Lisman JE, Grace AA. 2005. The hippocampal-VTA loop: Controlling the entry of information into long-term memory. *Neuron 46*: 703–713.

Lisman J, Coyle J, Green R, Javitt D, Benes F, Heckers S, Grace AA. 2008. Circuit-based framework for understanding neurotransmitter and risk gene interactions in schizophrenia. *Trends Neurosci 31*: 234–242.

Liu S, Schad DJ, Kuschpel MS, Rapp MA, Heinz A. 2016. Music and video gaming during breaks: Influence on habitual versus goal-directed decision making. *PLoS One 11*: e0150165.

Livingstone FB. 1993. On the nonexistence of human races. In: Harding S (Ed.), *The "Racial" Economy of Science*, pp. 133–141. Indiana University Press, Bloomington.

Logothetis NK, Pauls J, Augath M, Trinath T, Oeltermann A. 2001. Neurophysiological investigation of the basis of the fMRI signal. *Nature 412*: 150–157.

Lokwan SJ, Overton PG, Berry MS, Clark D. 2000. The medial prefrontal cortex plays an important role in the excitation of A10 dopaminergic neurons following intravenous muscimol administration. *Neuroscience 95*: 647–656.

Lubow RE, Gewirtz JC. 1995. Latent inhibition in humans: Data, theory, and implications for schizophrenia. *Psychol Bull 117*: 87–103.

Maia TV, Frank MJ. 2017. An integrative perspective on the role of dopamine in schizophrenia. *Biol Psychiatry 81*: 52–66.

Martel P, Fantino M. 1996. Influence of the amount of food ingested on mesolimbic dopaminergic system activity: A microdialysis study. *Pharmacol Biochem Behav 55*: 297–302.

Martinez D, Gil R, Slifstein M, Hwang DR, Huang Y, Perez A, Kegeles L, et al. 2005. Alcohol dependence is associated with blunted dopamine transmission in the ventral striatum. *Biol Psychiatry 58*: 779–786.

Martinez D, Saccone PA, Liu F, Slifstein M, Orlowska D, Grassetti A, Cook S, et al. 2012. Deficits in dopamine D(2) receptors and presynaptic dopamine in heroin dependence: Commonalities and differences with other types of addiction. *Biol Psychiatry 71*: 192–198.

Mataix-Cols D, Pertusa A, Snowdon J. 2011. Neuropsychological and neural correlates of hoarding: A practice-friendly review. *J Clin Psychol 67*: 467–476.

Mataix-Cols D, Wooderson S, Lawrence N, Brammer MJ, Speckens A, Phillips ML. 2004. Distinct neural correlates of washing, checking and hoarding symptom dimensions in obsessive-compulsive disorder. *Arch Gen Psychiatry 61*: 564–576.

Mayberg H. 1997. Limbic-cortical dysregulation: A proposed model of depression. *J Neuropsychiatry Clin Neurosci 9*: 471–481.

Mayberg HS, Liotti M, Brannan SK, McGinnis S, Mahurin RK, Jerabek PA, Silva JA, et al. 1999. Reciprocal limbic-cortical function and negative mood: Converging PET findings in depression and normal sadness. *Am J Psychiatry 156*: 675–682.

McCormick DA. 1992. Neurotransmitter actions in the thalamus and cerebral cortex. *J Clin Neurophysiol 9*: 212–223.

McCormick DA, Pape HC, Williamson A. 1992. Actions of norepinephrine in the cerebral cortex and thalamus: Implications for function of the central noradrenergic system. In: Barnes CD, Pompeiano O (Eds.), *Progress in Brain Research*, Vol. *88*, pp. 293–305. *Elsevier*, Amsterdam.

Mead GH. 1912. The mechanism of social consciousness. *J Philos Psychol Sci Methods 9*: 401–406.

Meehl PE. 1962. Schizotaxia, schizotypy, schizophrenia. *Am Psychol 17*: 827–838.

Méndez JE. 2014. Answer to Dr. Lieberman (President of the American Psychiatric Association) and Dr. Ruiz (President of the World Medical Association). In: *Torture in Healthcare Settings*, pp. 151–153. Center for Human Rights & Humanitarian Law, Washington College of Law, Washington. Available at http://antitorture.org/wp -content/uploads/2014/03/PDF_Torture_in_Healthcare_Publication.pdf.

Menzies L, Chamberlain SR, Laird AR, Thelen SM, Sahakian BJ, Bullmore ET. 2008. Integrating evidence from neuroimaging and neuropsychological studies of obsessive-compulsive disorder: The orbitofrontal-striatal model revisited. *Neurosci Biobehav Rev 32*: 525–549.

Meyer JS. 2013. 3,4-methylenedioxymethamphetamine (MDMA): Current perspectives. *Subst Abuse Rehabil 4*: 83–99.

Meyer-Lindenberg A. 2010. From maps to mechanisms through neuroimaging of schizophrenia. *Nature 468*: 194–202.

Misaki M, Suzuki H, Savitz J, Drevets WC, Bodurka J. 2016. Individual variations in nucleus accumbens responses associated with major depressive disorder subtypes. *Sci Rep 6*: 21227.

Miura T, Noma H, Furukawa TA, Mitsuyasu H, Tanaka S, Stockton S, Salanti G, et al. 2014. Comparative efficacy and tolerability of pharmacological treatments in the maintenance treatment of bipolar disorder: A systematic review and network meta-analysis. *Lancet Psychiatry 1*: 351–359.

Montague PR, Hyman SE, Cohen D. 2004. Computational roles for dopamine in behavioral control. *Nature 431*: 760–767.

Morel BA. 1857. *Traité des dégénérescences physiques, intellectuelles et morales de l'espèce humaine et des causes qui produisent ces variétés maladives.* J.B. Baillière, Paris.

Müller T, Büttner T, Kuhn W, Heinz A, Przuntek H. 1995. Palinopsia as sensory epileptic phenomenon. *Acta Neurol Scand 91*: 433–436.

Müller VI, Cieslik EC, Serbanescu I, Laird AR, Fox PT, Eickhoff SB. 2017. Altered brain activity in unipolar depression revisited. Meta-analysis of neuroimaging studies. *JAMA Psychiatry 74*: 47–55.

Munafo MR, Brown SM, Hariri AR. 2008. Serotonin transporter (5-HTTLPR) genotype and amygdale activation: A meta-analysis. *Biol Psychiatry 63*: 852–857.

Myrick H, Anton RF, Li X, Henderson S, Randall PK, Voronin K. 2008. Effect of naltrexone and ondansetron on alcohol cue-induced activation of the ventral striatum in alcohol-dependent people. *Arch Gen Psychiatry 65*: 466–475.

Nader K, Schafe GE, Le Doux JE. 2000. Fear memories require protein synthesis in the amygdala for reconsolidation after retrieval. *Nature 406*(6797): 722–726.

Northoff N, Bermpohl F. 2004. Cortical midline structures and the self. *Trends Cogn Sci 8*: 102–107.

O'Doherty JP, Dayan P, Friston K, Critchley H, Dolan RJ. 2003. Temporal difference models and reward-related learning in the human brain. *Neuron 38*: 329–337.

Oliveira FT, McDonald JJ, Goodman D. 2007. Performance monitoring in the anterior cingulate is not all error related: Expectancy deviation and the representation of action-outcome associations. *J Cogn Neurosci 19*: 1994–2004.

Otto AR, Gershman SJ, Markman AB, Daw ND. 2013a. The curse of planning: Dissecting multiple reinforcement-learning systems by taxing the central executive. *Psychol Sci 24*: 751–761.

Otto AR, Raio CM, Chiang A, Phelps EA, Daw ND. 2013b. Working-memory capacity protects model-based learning from stress. *Proc Natl Acad Sci USA 110*: 20941–20946.

Otto AR, Skatova A, Madlon-Kay S, Daw ND. 2015. Cognitive control predicts use of model-based reinforcement learning. *J Cogn Neurosci 27*: 319–333.

Pankow A, Friedel E, Sterzer P, Seiferth N, Walter H, Heinz A, Schlagenhauf F. 2013. Altered amygdala activation in schizophrenia patients during emotion processing. *Schizophr Res 150*: 101–106.

Pankow A, Deserno L, Walter M, Fydrich T, Bermpohl F, Schlagenhauf F, Heinz A. 2015. Reduced default mode network connectivity in schizophrenia patients. *Schizophr Res 165*: 90–93.

Pankow A, Katthagen T, Diner S, Deserno L, Boehme R, Kathmann N, Gleich T, et al. 2016. Aberrant salience is related to dysfunctional self-referential processing in psychosis. *Schizophr Bull 42*: 67–76.

Park SQ, Kahnt T, Beck A, Cohen MX, Dolan RJ, Wrase J, Heinz A. 2010. Prefrontal cortex fails to learn from reward prediction errors in alcohol dependence. *J Neurosci 30*: 7749–7753.

Parkinson JA, Robbins TW, Everitt BJ. 2000. Dissociable roles of the central and basolateral amygdala in appetitive emotional learning. *Eur J Neurosci 12*: 405–413.

Pauen M. 2010. How privileged is first-person privileged access? *Am Philos Q 47*: 1–15.

Paulus MP, Tapert SF, Schuckit MA. 2005. Neural activation patterns of methamphetamine-dependent subjects during decision making predict relapse. *Arch Gen Psychiatry 62*: 761–768.

Pavlov IP. 1928. *Lectures on Conditioned Reflexes*. Allen and Unwin, London.

Pezawas L, Meyer-Lindenberg A, Dabant EM, Verchinski BA, Munoz KE, Kolachana BS, Egan F, et al. 2005. 5-HTTLPR polymorphism impacts human cingulate-amygdala interactions: A genetic susceptibility mechanism for depression. *Nat Neurosci 8*: 828–834.

Pfefferbaum A, Desmond JE, Galloway C, Menon V, Glover GH, Sullivan EV. 2001. Reorganization of frontal systems used by alcoholics for spatial working memory: An fMRI study. *Neuroimage 14*: 7–20.

Plessner H. 1975. *Die Stufen des Organischen und der Mensch*. Walter de Gruyter, Berlin, New York.

Plessner H. 2003a. *Über den Begriff der Leidenschaft*. In: Plessner H. *Gesammelte Schriften in 10 Bänden, Vol. VIII: Conditio humana*, pp. 66–76. Suhrkamp, Frankfurt/M.

Plessner H. 2003b. *Der kategorische Konjunktiv. Ein Versuch über die Leidenschaft*. In: Plessner H. *Gesammelte Schriften in 10 Bänden, Bd. VIII: Conditio humana*, pp. 338–352. Suhrkamp, Frankfurt/M.

Prevost C, Liljeholm M, Tyszka JM, O'Doherty JP. 2012. Neural correlates of specific and general Pavlovian-to-instrumental transfer within human amygdalar subregions: A high resolution fMRI study. *J Neurosci 32*: 8383–8390.

Price JL, Drevets WC. 2012. Neural circuits underlying the pathophysiology of mood disorders. *Trends Cogn Sci 16*: 61–71.

Puls I, Mohr J, Wrase J, Vollstädt-Klein S, Leménager T, Vollmert C, Rapp M, et al. 2009. A model comparison of COMT effects on central processing of affective stimuli. *Neuroimage 46*(3): 683.

Pycock CJ, Kerwin RW, Carter CJ. 1980. Effect of lesion of cortical dopamine terminals on subcortical dopamine receptors in rats. *Nature 286*: 74–76.

Radenbach C, Reiter AM, Engert V, Sjoerds Z, Villringer A, Heinze HJ, Deserno L, Schlagenhauf F. 2015. The interaction of acute and chronic stress impairs model-based behavioral control. *Psychoneuroendocrinol 53*: 268–280.

Rado S. 1956. *Psychoanalysis of Behavior*. Grune and Stratton, New York.

Rangel A, Camerer C, Montague PR. 2008. A framework for studying the neurobiology of value-based decision making. *Nat Rev Neurosci 9*: 545–556.

Rapp MA, Kluge U, Penka S, Vardar A, Aichberger MC, Mundt AP, Schouler-Ocak M, et al. 2015. When local poverty is more important than your income: Mental health in minorities in inner cities. *World Psychiatry 14*(2): 249–250.

Redish AD. 2004. Addiction as a computational process gone awry. *Science 306*: 1944–1947.

Redish AD, Johnson A. 2007. A computational model of craving and obsession. *Ann N Y Acad Sci 1104*: 324–339.

Reimold M, Batra A, Knobel A, Smolka MN, Zimmer A, Mann K, Solbach C, et al. 2008. Anxiety is associated with reduced serotonin transporter availability in unmedicated patients with unipolar depression: A [11C]DASB PET study. *Mol Psychiatry 13*: 606–613.

Reimold M, Knobel A, Rapp M, Batra A, Wiedemann K, Ströhle A, Zimmer A, et al. 2010. Central serotonin transporter levels are associated with stress hormone response and anxiety. *Psychopharmacology (Berl) 213*: 563–572.

Rescorla RA, Wagner AR. 1972. A theory of Pavlovian conditioning: Variations in the effectiveness of reinforcement on nonreinforcement. In: Black AH, Prokasy WF (Eds.), *Classical Conditioning 2: Current Research and Theory*, pp. 64–99. Appleton-Century-Crofts, New York.

Reuter J, Raedler T, Rose M, Hand I, Gläscher J, Büchel C. 2005. Pathological gambling is linked to reduced activation of the mesolimbic reward system. *Nat Neurosci 8*: 147–148.

Rinne JO, Kuikka JT, Bergström KA, Rinne UK. 1995. Striatal dopamine transporter in different disability stages of Parkinson's disease studied with [I-123]β-CIT SPECT. *Park Rel Dis 1*: 47–51.

Risch N, Herrell R, Lehner T, Liang KY, Eaves L, Hoh J, Griem A, et al. 2009. Interaction between the serotonin transporter gene (5-HTTLPR), stressful life events, and risk of depression: A meta-analysis. *JAMA 30*: 2462–2471.

Robbins TW, Everitt BJ. 1996. Neurobehavioral mechanisms of reward and motivation. *Curr Opin Neurobiol 6*: 228–236.

Robinson TE, Berridge KC. 1993. The neural basis of drug craving: An incentive-sensitization theory of addiction. *Brain Res Brain Res Rev 18*: 247–291.

Robinson OJ, Cools R, Sahakian BJ. 2012. Tryptophan depletion disinhibits punishment but not reward prediction: Implications for resilience. *Psychopharmacology (Berl) 219*: 599–605.

Rolls E, Loh M, Deco G, Winterer G. 2008. Computational models of schizophrenia and dopamine modulation in the prefrontal cortex. *Nat Rev Neurosci 9*: 696–709.

Romanczuk-Seiferth N, Koehler S, Dreesen C, Wüstenberg T, Heinz A. 2015. Pathological gambling and alcohol dependence: Neural disturbances in reward and loss avoidance processing. *Addict Biol 20*: 557–569.

Roman-Vendrell C, Yudowski GA. 2015. Real-time imaging of mu-opioid receptors by internal reflection fluorescence microscopy. *Methods Mol Biol 1230*: 79–86.

Rommelspacher H, Raeder C, Kaulen P, Brüning G. 1992. Adaptive changes of dopamine-D2 receptors in rat brain following ethanol withdrawal: A quantitative autoradiographic investigation. *Alcohol 9*: 355–362.

Rüdin E. 1939. Eröffnungsansprache zur V. Jahresversammlung der Gesellschaft Deutscher Neurologen und Psychiater. *Zeitschrift ges Neurol Psychiatrie 81*: 164–167.

Rupp CI, Beck JK, Heinz A, Kemmler G, Manz S, Tempel K, Fleischhacker WW. 2016. Impulsivity and alcohol dependence treatment completion: Is there a neurocognitive risk factor at treatment entry? *Alcohol Clin Exp Res 40*: 152–160.

Russel JA, Weiss A, Mendelsohn GA. 1989. The affective grid: A single-item scale of pleasure and arousal. *J Pers Soc Psychol 57*: 493–502.

Rygula R, Clarke HF, Cardinal RN, Cockcroft GJ, Xia J, Dalley JW, Robbins TW, Roberts AC. 2015. Role of central serotonin in anticipation of rewarding and punishing outcomes: Effects of selective amygdala or orbitofrontal 5-HT depletion. *Cereb Cortex 25*: 3064–3076.

Sadacca BF, Jones JL, Schoenbaum G. 2016. Midbrain dopamine neurons compute inferred and cached value prediction errors in a common framework. *eLife 5*: pii e13665.

Santarelli L, Saxe M, Gross C, Surget A, Battaglia F, Dulawa S, Weisstaub N, et al. 2003. Requirement of hippocampal neurogenesis for the behavioral effects of antidepressants. *Science 301*: 805–809.

Sartorius N. 2010. Meta-effects of classifying mental disorders. In: Regier DA, Narrow WE, Kuhl GA, Kupfer DJ (Eds.), *The Conceptual Evolution of DSM-5.*, pp. 59–77. American Psychiatric Association, Arlington, VA.

Sartorius N, Jablensky A, Korten A, Ernberg G, Anker M, Cooper JE, Day R. 1986. Early manifestation and first-contact incidence of schizophrenia in different cultures. A preliminary report on the initial evaluation phase of the WHO Collaborative Study on determinants of outcome of severe mental disorders. *Psychol Med 16*: 909–928.

Sartre JP. 1943. *L'être et le néant. Essai d'ontologie phénoménologique*. Gallimard, Paris. (Reprinted 2005)

Saunders RC, Kolachana BS, Bachevalier J, Weinberger DR. 1998. Neonatal lesions of the medial temporal lobe disrupt prefrontal cortical regulation of striatal dopamine. *Nature 393*: 169–171.

Saxena S, Brody AL, Maidment KM, Dunkin JJ, Colgan M, Alborzian S, Phleps ME, Baxter LR, Jr. 1999. Localized orbitofrontal and subcortical metabolic changes and predictors of response to paroxetine treatment in obsessive-compulsive disorder. *Neuropsychopharmacology 21*: 683–693.

Schad DJ, Jünger E, Sebold M, Garbusow M, Bernhardt N, Javadi AH, Zimmermann US, et al. 2014. Processing speed enhances model-based over model-free reinforcement learning in the presence of high working memory functioning. *Front Psychol 5*: 1450.

Schizophrenia Working Group of the Psychiatric Genomics Consortium. 2014. Biological insights from 108 schizophrenia-associated genetic loci. *Nature 511*(7510): 421–427.

Schlagenhauf F, Sterzer P, Schmack K, Ballmaier M, Rapp M, Wrase J, Juckel G, et al. 2009. Reward feedback alterations in unmedicated schizophrenia patients: Relevance for delusions. *Biol Psychiatry 65*: 1032–1039.

Schlagenhauf F, Rapp MA, Huys QJ, Beck A, Wüstenberg T, Deserno L, Buchholz HG, et al. 2013. Ventral striatal prediction error signaling is associated with dopamine synthesis capacity and fluid intelligence. *Hum Brain Mapp 34*: 1490–1499.

Schlagenhauf F, Huys QJ, Deserno L, Rapp MA, Beck A, Heinze HJ, Dolan R, Heinz A. 2014. Striatal dysfunction during reversal learning in unmedicated schizophrenia patients. *Neuroimage 89*: 171–180.

Schmack K, Gòmez-Carrillo de Castro A, Rothkirch M, Sekutowicz M, Rössler H, Haynes JD, Heinz A, et al. 2013. Delusions and the role of beliefs in perceptual inference. *J Neurosci 33*: 13701–13712.

Schmidt K, Roiser JP. 2009. Assessing the construct validity of aberrant salience. *Front Behav Neurosci 3*: 58.

Schmidt K, Nolte-Zenker B, Patzer J, Bauer M, Schmidt LG, Heinz A. 2001. Psychopathological correlates of reduced dopamine receptor sensitivity in depression, schizophrenia, and opiate and alcohol dependence. *Pharmacopsychiatry 34*: 66–72.

Schneider K. 1942. *Psychischer Befund und psychiatrische Diagnose*. Thieme, Leipzig.

Schneider K. 1967. *Psychopathologie*. Thieme, Stuttgart.

Schoofs N, Heinz A. 2013. Pathological gambling. Impulse control disorder, addiction or compulsion? *Nervenarzt 84*: 629–634.

Schott BH, Minuzzi L, Krebs RM, Elmenhorst D, Lang M, Winz OH, Seidenbecher CI, et al. 2008. Mesolimbic functional magnetic resonance imaging activations during reward anticipation correlate with reward-related ventral striatal dopamine release. *J Neurosci 28*: 14311.

Schott BH, Voss M, Wagner B, Wüstenberg T, Düzel E, Behr J. 2015. Fronto-limbic novelty processing in acute psychosis: Disrupted relationship with memory performance and potential implications for delusions. *Front Behav Neurosci 9*: 144.

Schuckit MA, Smith TL. 1996. An 8-year follow-up of 450 sons of alcoholic and control subjects. *Arch Gen Psychiatry 53*: 202–210.

Schultz W, Dayan P, Montague PR. 1997. A neural substrate of prediction and reward. *Science 275*: 1593–1599.

Schultz W. 2004. Neural coding of basic reward terms of animal learning theory, game theory, microeconomics and behavioral ecology. *Neurobiology 14*: 139–147.

Schultz W. 2007. Multiple dopamine functions at different time courses. *Annu Rev Neurosci 30*: 259–288.

Sebold M, Deserno L, Nebe S, Schad DJ, Garbusow M, Hägele C, Keller J, et al. 2014. Model-based and model-free decisions in alcohol dependence. *Neuropsychobiology 70*: 122–131.

Sebold M, Schad DJ, Nebe S, Garbusow M, Jünger E, Kroemer NB, Kathmann N, et al. 2016. Don't think, just feel the music: Individuals with strong Pavlovian-to-instrumental transfer effects rely less on model-based reinforcement learning. *J Cogn Neurosci 4*: 1–11.

Sen S, Burmeister M, Ghosh D. 2004. Meta-analysis of the association between a serotonin transporter promoter polymorphism (5-HTTLPR) and anxiety-related personality traits. *Am J Med Genet B Neuropsychiatr Genet 127B*(1): 85–89.

Seo S, Mohr J, Beck A, Wüstenberg T, Heinz A, Obermayer K. 2015. Predicting the future relapse of alcohol-dependent patients from structural and functional brain images. *Addict Biol 20*: 1042–1055.

Shaham Y, Shalev U, Lu L, De Wit H, Stewart J. 2003. The reinstatement model of drug relapse: History, methodology and major findings. *Psychopharmacology (Berl) 168*: 3–20.

Sharp ME, Foerde K, Daw ND, Shohamy D. 2016. Dopamine selectively remediates "model-based" reward learning: A computational approach. *Brain 139*: 355–364.

Sharpley CF, Palanisamy SK, Glyde NS, Dillingham PW, Agnew LL. 2014. An update on the interaction between the serotonin transporter promoter variant (5-HTTLPR), stress and depression, plus an exploration of non-confirming findings. *Behav Brain Res 273*: 89–105.

Shoemaker S. 1996. *The First-Person Perspective and Other Essays*. Cambridge University Press, New York.

Shumay E, Folwer JS, Wang GJ, Logan J, Alia-Klein N, Goldstein RZ, Maloney T, et al. 2012. Repeat variation in the human PER2 gene as a new genetic marker associated with cocaine addiction and brain dopamine D2 receptor availability. *Transl Psychiatry 2*: e86.

Sjoerds Z, de Wit S, van den Brink W, Robbins TW, Beekman ATF, Penninx BWJH, Veltman DJ. 2013. Behavioral and neuroimaging evidence for overreliance on habit learning in alcohol-dependent patients. *Transl Psychiatry 3*: e337.

Skinner BF. 1953. *Science and Human Behavior*. Macmillan, New York.

Slifstein M, van de Giessen E, Van Snellenberg J, Thompson JL, Narendran R, Gil R, Hackett E, et al. 2015. Deficits in prefrontal cortical and extrastriatal dopamine release in schizophrenia: A positron emission tomographic functional magnetic resonance imaging study. *JAMA Psychiatry 72*: 316–324.

Smith TD, Kuczenski R, George-Friedman K, Malley JD, Foote SL. 2000. In vivo microdialysis assessment of extracellular serotonin and dopamine levels in awake monkeys during sustained fluvoxamine administration. *Synapse 38*: 460–470.

Smolka MN, Schumann G, Wrase J, Grüsser SM, Flor H, Mann K, Braus DF, et al. 2005. Catechol-O-methyltransferase val158met genotype affects processing of emotional stimuli in the amygdala and prefrontal cortex. *J Neurosci 25*(4): 836–842.

Smolka MN, Bühler M, Schumann G, Klein S, Hu XZ, Moayer M, Zimmer A, et al. 2007. Gene-gene effects on central processing of aversive stimuli. *Mol Psychiatry 12*(3): 307–317.

Soares-Weiser K, Maayan N, Bergman H, Davenport C, Kirkham AJ, Grabowski S, Adams CE. 2015. First rank symptoms for schizophrenia (Cochrane Diagnostic Test Accuracy Review). *Schizophr Bull 41*: 792–794.

Sora I, Hall FS, Andrews AM, Itokawa M, Li XF, Wei HB, Wichems C, et al. 2001. Molecular mechanisms of cocaine reward: Combined dopamine and serotonin transporter knockouts eliminate cocaine place preference. *Proc Natl Acad Sci USA 98*: 5300–5305.

Stephan KE, Schlagenhauf S, Huys QJ, Raman S, Aponte EA, Brodersen KH, Rigoux L, et al. 2016. Computational neuroimaging strategies for single patient prediction. *Neuroimage 145*: 180–199.

Stetler C, Miller GE. 2011. Depression and hypothalamic-pituitary-adrenal axis: A quantitative summary of four decades of research. *Psychosom Med 73*: 114–126.

Stoy M, Schlagenhauf F, Sterzer P, Bermpohl F, Hägele C, Suchotzki K, Schmack K, et al. 2012. Hyporeactivity of ventral striatum towards incentive stimuli in unmedicated depressed patients normalizes after treatment with escitalopram. *J Psychopharmacol 26*: 677–688.

Streifler MB, Korczyn AD, Melamed E, Youdim MBH. 1990. *Parkinson's Disease: Anatomy, Pathology, and Therapy*. Advances in Neurology, Vol. 53. Raven Press, New York.

Stuke H, Gutwinski S, Wiers CE, Schmidt TT, Gröpper S, Parnack J, Gawron C, et al. 2016. To drink or not to drink: Harmful drinking is associated with hyperactivation of reward areas rather than hypoactivation of control areas in men. *J Psychiatry Neurosci 41*: 24–36.

Sutton R, Barto A. 1998. *Reinforcement Learning: An Introduction*. MIT Press, Cambridge, MA.

Swartz JR, Hariri AR, Williamson DE. 2017. An epigenetic mechanism links socioeconomic status to changes in depression-related brain function in high-risk adolescents. *Mol Psychiatry 22*: 209–214.

Szasz T. 1970. *Ideology and Insanity: Essays on the Psychiatric Dehumanization of Man*. Syracuse University Press, Syracuse, NY.

Taber MT, Das S, Fibiger HC. 1995. Cortical regulation of dopamine release: Mediation via the ventral tegmental area. *J Neurochem 65*: 1407–1410.

Tecott LH, Julius D. 1993. A new wave of serotonin receptors. *Curr Opin Neurobiol 3*: 310–315.

Thanos PK, Michaelidis M, Umegaki H, Volkow ND. 2008. D2R DNA transfer into the nucleus accumbens attenuates cocaine self-administration in rats. *Synapse 62*: 481–486.

Tiffany ST. 1999. Cognitive concepts of craving. *Alcohol Res Health 23*: 215–224.

Tiffany ST, Carter BL. 1998. Is craving the source of compulsive drug use? *J Psychopharmacol 12*: 23–30.

Tishkoff SA, Dietzsch E, Speed W, Pakstis AJ, Kidd JR, Cheung K, Bonné-Tamir B, et al. 1996. Global patterns of linkage disequilibrium at the CD4 locus and modern human origins. *Science 271*: 1380–1387.

Toller E. 1933/2010. *Eine Jugend in Deutschland*. Anaconda Verlag, Köln.

Tremblay LK, Naranjo CA, Graham SJ, Hermann N, Mayberg H, Hevenor S, Busto UE. 2005. Functional neuroanatomical substrates of altered reward processing in major depressive disorder revealed by a dopaminergic probe. *Arch Gen Psychiatry 62*: 1228–1236.

Tsai G, Gastfriend DR, Coyle JT. 1995. The glutamatergic basis of human alcoholism. *Am J Psychiatry 152*: 332–340.

United Nations. 2006. Convention on the Rights of Persons with Disabilities. Available at https://www.un.org/development/desa/disabilities/convention-on-the-rights -of-persons-with-disabilities.html [last access: 31.03.2017]

van Os J. 2016. "Schizophrenia" does not exist. *BMJ 352*: i375.

van Praag HM. 1967. The possible significance for cerebral dopamine for neurology and psychiatry. *Psychiatr Neurol Neurochir 70*: 361–379.

van Praag HM, Asnis GM, Kahn RS, Brown SL, Korn M, Friedman JM, Wetzler S. 1990. Nosological tunnel vision in biological psychiatry; A plea for a functional psychopathology. *Ann N Y Acad Sci 600*: 501–510.

Veling W. 2013. Ethnic minority position and risk for psychotic disorders. *Curr Opin Psychiatry 26*: 166–171.

Virkkunen M, Kallio E, Rawlings R, Tokola R, Poland RE, Guidotti A, Nemeroff C, et al. 1994. Personality profiles and state aggressiveness in Finnish alcoholic, violent offenders, fire setters, and healthy volunteers. *Arch Gen Psychiatry 51*: 28–33.

Volkow ND, Fowler JS. 2000. Addiction, a disease of compulsion and drive: Involvement of the orbitofrontal cortex. *Cereb Cortex 10*(2): 318–325.

Volkow ND, Hitzemann R, Wang GJ, Fowler JS, Wolf AP, Dewey SL, Handlesman L. 1992. Long-term frontal brain metabolic changes in cocaine abusers. *Synapse 11*: 184–190.

Volkow ND, Wang GJ, Fowler JS, Logan J, Hitzemann R, Ding YS, Pappas N, et al. 1996. Decreases in dopamine receptors but not in dopamine transporters in alcoholics. *Alcohol Clin Exp Res 20*: 1594–1598.

Volkow ND, Wang GJ, Fowler JS, Logan J, Gately SJ, Hitzemann R, Chen AD, et al. 1997. Decreased striatal dopaminergic responsiveness in detoxified cocaine-dependent subjects. *Nature 386*: 830–833.

Volkow ND, Chang L, Wang GJ, Fowler JS, Ding YS, Sedler M, Logan J, et al. 2001. Low level of brain dopamine D2 receptors in methamphetamine abusers: Association with metabolism in the orbitofrontal cortex. *Am J Psychiatry 158*: 2015–2021.

Volkow ND, Wang GJ, Begleiter H, Porjesz B, Fowler JS, Telang F, Wong C, et al. 2006. High levels of dopamine D2 receptors in unaffected members of alcoholic families: Possible protective factors. *Arch Gen Psychiatry 63*: 999–1008.

Volkow ND, Tomasi D, Wang GJ, Logan J, Alexoff DL, Javne M, Fowler JS, et al. 2014. Stimulant-induced dopamine increases are markedly blunted in active cocaine abusers. *Mol Psychiatry 19*: 1037–1043.

Volkow ND, Wang GJ, Shokri Kojori E, Fowler JS, Benveniste H, Tomasi D. 2015a. Alcohol decreases baseline brain glucose metabolism more in heavy drinkers than controls but has no effect on stimulation-induced metabolic increases. *J Neurosci 35*: 3248–3255.

Volkow ND, Wang GJ, Smith L, Fowler JS, Telang F, Logan J, Tomasi D. 2015b. Recovery of dopamine transporters is not linked to changes in dopamine release. *Neuroimage 121*: 20–28.

Vollstädt-Klein S, Wichert S, Rabinstein J, Bühler M, Klein O, Ende G, Hermann D, Mann K. 2010. Initial, habitual and compulsive alcohol use is characterized by a shift of cue processing from ventral to dorsal striatum. *Addiction 105*: 1741–1749.

Voon V, Pessiglione M, Brezing C, Gallea C, Fernandez HH, Dolan RJ, Hallett M. 2010. Mechanisms underlying dopamine-mediated reward bias in compulsive behaviors. *Neuron 65*: 135–142.

Voon V, Derbyshire K, Rück C, Irvine MA, Worbe Y, Enander J, Schreiber LR, et al. 2015. Disorders of compulsivity: A common bias towards learning habits. *Mol Psychiatry 20*: 345–352.

Voss M, Moore J, Hauser M, Gallinat J, Heinz A, Haggard P. 2010. Altered awareness of action in schizophrenia: A specific deficit in predicting action consequences. *Brain 133*: 3104–3112.

Wang XJ, Krystal J. 2014. Computational psychiatry. *Neuron 84*: 638–654.

Wassum KM, Ostlund SB, Loewinger GC, Maidment NT. 2013. Phasic mesolimbic dopamine release tracks reward seeking during expression of Pavlovian-to-instrumental transfer. *Biol Psychiatry 73*: 747–755.

Watson D, Clark LA, Tellegen A. 1988. Development and validation of brief measures of positive and negative affect: The PANAS scales. *J Pers Soc Psychol 54*: 1063–1070.

Weinberger DR. 1987. Implications of normal brain development for the pathogenesis of schizophrenia. *Arch Gen Psychiatry 44*: 660–669.

Weinberger DR. 1996. On the plausibility of "the neurodevelopmental hypothesis" of schizophrenia. *Neuropsychopharmacology 14*: 1–11.

Weinstein JJ, Chochan MO, Slifstein M. Kegeles LS, Moore H, Abi-Dargham A. 2016. Pathway-specific dopamine abnormalities in schizophrenia. *Biol Psychiatry 81*: 31–42.

Whelan R, Watts R, Orr CA, Althoff RR, Artiges E, Banaschewski T, Barker GJ, et al. 2014. Neuropsychological profiles of current and future adolescent alcohol misusers. *Nature 512*: 185–189.

Wiers RW, Eberl C, Rinck M, Becker ES, Lindenmeyer J. 2011. Retraining automatic action tendencies changes alcoholic patients' approach bias for alcohol and improves treatment outcome. *Psychol Sci 22*: 490–497.

Wiers CE, Stelzel C, Park QS, Gawron CK, Ludwig VU, Gutwinski S, Heinz A, et al. 2014. Neural correlates of alcohol-approach bias in alcohol addiction: The spirit is willing but the flesh is weak for spirits. *Neuropsychopharmacology 39*: 688–697.

Williams GV, Goldman-Rakic PS. 1995. Modulation of memory fields by dopamine D1 receptors in prefrontal cortex. *Nature 376*: 572–575.

Winterer G, Ziller M, Dorn H, Frick K, Mlert C, Wuebben Y, Herrmann WM, Coppola R. 2000. Schizophrenia: Reduced signal-to-noise ratio during information processing. *Clin Neurophysiol 111*: 837–849.

Winterer G, Coppola R, Goldberg TE, Egan MF, Jones DW, Sanchez CE, Weinberger DR. 2004. Prefrontal broadband noise, working memory, and genetic risk for schizophrenia. *Am J Psychiatry 161*: 490–500.

Winterer G, Musso F, Beckmann C, Mattay V, Egan MF, Jones DW, Callicott JH, et al. 2006. Instability of prefrontal signal processing in schizophrenia. *Am J Psychiatry 163*: 1960–1968.

Wise RA. 1982. Neuroleptics and operant behavior: The anhedonia hypothesis. *Behav Brain Sci 5*: 39–87.

Wise RA. 1985. The anhedonia hypothesis: Mark III. *Behav Brain Sci 8*: 178–186.

Wise RA. 1988. The neurobiology of craving: Implications for the understanding and treatment of addiction. *J Abnorm Psychol 97*: 118–132.

Wolf SS, Jones DW, Knable MB, Gorey J, Lee KS, Hyde TM, Coppola R, Weinberger DR. 1996. Tourette syndrome: prediction of phenotypic variation in monozygotic twins by caudate nucleus D2 receptor binding. *Science 273*: 1225–1227.

Worbe Y, Palminteri S, Savulich G, Daw ND, Fernandez-Egea E, Robbins TW, Voon V. 2015. Valence-dependent influence of serotonin depletion on model-based choice strategy. *Mol Psychiatry 21*: 624–629.

World Health Organization. 2011. *ICD-10, International Statistical Classification of Diseases and Related Health Problems*, 10th Revision. WHO, Geneva.

Wrase J, Schlagenhauf F, Kienast T, Wüstenberg T, Bermpohl F, Kahnt T, Beck A, et al. 2007. Dysfunction of reward processing correlates with alcohol craving in detoxified alcoholics. *Neuroimage 35*: 787–794.

Wunderink L, Nieboer RM, Wiersma D, Sytema S, Nienhuis FJ. 2013. Recovery in remitted first-episode psychosis at 7 years of follow-up of an early dose of reduction/ discontinuation or maintenance treatment strategy: long-term follow-up of a 2-year randomized clinical trial. *JAMA Psychiatr 70*: 913–920.

Yacubian J, Sommer T, Schroeder K, Gläscher J, Kalisch R, Leuenberger B, Braus DF, Büchel C. 2007. Gene-gene interaction associated with neural reward sensitivity. *Proc Natl Acad Sci USA 104*(19): 8125–8130.

Yuii K, Suzuki M, Kurachi M. 2007. Stress sensitization in schizophrenia. *Ann N Y Acad Sci 1113*: 276–290.

Zangen A, Nakash R, Yadid G. 1999. Serotonin-mediated increases in the extracellular levels of beta-endorphin in the arcuate nucleus and nucleus accumbens: A microdialysis study. *J Neurochem 73*: 2569–2574.

Zubieta JK, Stohler CS. 2009. Neurobiological mechanisms of placebo response. *Ann N Y Acad Sci 1156*: 198–201.

Name Index

Abercrombie, E. D., 98
Abi-Dargham, A., 50, 97, 101, 106
Adams, K. M., 128, 140
Adams, R. A., 51, 104, 106, 107, 162
Akam, T., 39
Alexander, G. E., 48, 51, 65
American Association of Physical
    Anthropologists (AAPA), 69
American Psychiatric Association, 5, 75,
    81, 84, 113, 115, 116, 143, 146, 147,
    149, 164, 168
Anderson, G. M., 55, 58
Andreasen, N. C., 82
Asensio, S., 119
Atzil, A., 108

Bandelow, B., 129
Barch, D. M., 81
Bart, G., 121
Bauer, M., 147
Baxter, L. R., 128, 140, 172
Bayes, T., 104
Beck, A., 124, 135, 136, 138
Beesdo, K., 153
Bel, N., 58
Belin, D., 119
Belin-Rauscent, A., 129
Benmansour, S., 151
Bermpohl, F., 160, 161
Berridge, K. C., 30, 32, 97, 159
Bienvenu, O. J., 166

Blankenburg ,W., 94, 95, 108
Bleuler, E., 70, 76–83, 94, 95,
    109, 164
Blumenbach, J. F., 68, 69
Bobon, D. P., 93, 148, 149
Bock, T., 75
Boehme, R., 99, 102
Bohman, M., 123
Boorse, C., 7
Braus, D. F., 124
Brodie, M. S., 55
Brody, A. L., 128
Büchel, C., 17, 60, 150

Callicott, J. H., 83
Cantor-Graae, E., 110
Carlsson, A., 158
Carpenter, W. T., 81, 82
Carter, C. S., 124
Cartoni, E., 24
Caspi, A., 123, 152, 153, 166
Chamorro, A. J., 121
Charlet, K., 133
Clarke, H. F., 62
Cloninger, C. R., 119
Cohen, J. D., 50
Conrad, K., 98
Cools, R., 50, 165
Corbit, J. H., 24
Corbit, L. H., 24
Crawford, J. R., 4

Subject Index